Nanowires - Recent Progress

Edited by Xihong Peng

Published in London, United Kingdom

IntechOpen

Supporting open minds since 2005

Nanowires - Recent Progress
http://dx.doi.org/10.5772/intechopen.87902
Edited by Xihong Peng

Contributors
M.Ajmal Khan, Yasuaki Ishikawa, Fangliang Gao, Qing Liu, Jiang Shi, Shuti Li, Soumyaranjan Routray, Trupti Ranjan Lenka, Jun-Sik Yoon, Jinsu Jeong, Seunghwan Lee, Junjong Lee, Rock-Hyun Baek, Chinnamuthu Paulsamy, Pheiroijam Pooja, Heigrujam Manas Singh, Nonofo Mathiba Jack Ditshego, Ricardo Tellez-Limon, Rafael Salas-Montiel, Andreea Costas, Nicoleta Preda, Camelia Florica, Ionut Enculescu, mohamed mbarek, Alimi Kamel, Carina B. Maliakkal

First published in London, United Kingdom, 2021 by IntechOpen
IntechOpen is the global imprint of INTECHOPEN LIMITED, registered in England and Wales, registration number: 11086078, 5 Princes Gate Court, London, SW7 2QJ, United Kingdom
Printed in Croatia

British Library Cataloguing-in-Publication Data
A catalogue record for this book is available from the British Library

Additional hard and PDF copies can be obtained from orders@intechopen.com

Nanowires - Recent Progress
Edited by Xihong Peng
p. cm.
Print ISBN 978-1-83962-391-2
Online ISBN 978-1-83962-392-9
eBook (PDF) ISBN 978-1-83962-393-6

For EU product safety concerns:
IN TECH d.o.o., Prolaz Marije Krucifikse Kozulić 3, 51000 Rijeka, Croatia, info@intechopen.com or visit our website at intechopen.com.

Meet the editor

Dr. Xihong Peng is a professor in the College of Integrative Sciences and Arts, Arizona State University (ASU), Polytechnic Campus. She received a Ph.D. in Physics from Rensselaer Polytechnic Institute, New York, in 2007. She joined ASU as an assistant professor in 2008 and was promoted to professor in 2021. Dr. Peng's research interests are to explore novel materials and seek their applications in nanoelectronics and alternative energies, as well as to gain a fundamental understanding of the materials' properties at an atomic level. She has applied physics-based computational methods to investigate a wide range of materials, including group IV, III-V semiconductor nanostructures, and novel materials for applications in fuel cells, photocatalysts, and Li-ion batteries.

Contents

Preface

Quasi-one-dimensional nanoscale materials such as nanowires have been a research focus in the fields of material sciences and engineering in the last few decades. A series of books emphasizing the research areas of nanowires have been published by several scientific publishers. This book provides a review of the recent progress of nanowires and their wide range of applications, as well as the associated advancements in material synthesis and characterization.

Low-dimensional nanoscale materials exhibit promising applications in today's science and technology, due to their reduction in size, large surface-to-volume ratio, and novel properties resulting from quantum confinement effects. The significantly larger surface-to-volume ratio in nanostructures, compared with their bulk counterparts, leads to high sensitivity to surface effects and enables a new generation of technologies in research fields. Compared to other low-dimensional structures, such as zero-dimensional quantum dots (or nanoclusters) and two-dimensional nanosheets, one-dimensional nanowires demonstrate unique geometrical advantages that expedite the application of nanowires as bases of electronic devices, such as channels and interconnects. In such applications, the reduction in size of nanowires enables a faster speed and a greater power density, in addition to a reduced device form factor. Numerous conventional techniques established in traditional bulk devices are readily applicable to nanowire devices, which is expected to boost extensive applications of nanowires.

This book contains ten chapters divided into three sections: Oxide Nanowires, Group III–V Compounds, and Other Nanowires.

Section 1 examines recent progress in metal oxide nanowires, which include novel transition metal oxide nanowire field-effect transistors for biosensing, nanowires as building blocks for optoelectronic devices, and an advanced technique of glancing angle deposition for nanowire synthesis and their applications. Transition metal oxides, such as ZnO, CuO, TiO_2, SnO_2, and WO_3, possess various advanced properties, for instance, resilient catalyst properties and large magnetoresistance coefficients. These exceptional properties are fundamentally correlated to the unfilled d-electrons in the transition metals. Transition metal oxide nanowires, combined with their unique size, geometrical effects, and stoichiometry engineering, are expected to reveal novel properties/applications and to play an important role in many different fields of science and technology. Chapter 1 provides a review of ZnO nanowire field-effect transistors and biosensors, emphasizing the different ways to improve the properties and performance of doped ZnO as a channel material. It suggests that top-down fabrication processes are preferred over bottom-up ones due to the former's enhanced flexibility of geometrical dimension control and capability of mass production. Chapter 2 focuses on two types of metal oxide nanowire arrays: the n-type eco-friendly and versatile ZnO and p-type highly stable CuO, using two straightforward and cost-effective preparation wet and dry methods of chemical synthesis in aqueous solution and thermal oxidation in air, respectively. Electronic devices based on single metal oxide nanowires were developed and analyzed in terms of electrical characterization. Chapter 3 gives reviews the synthesis of metal oxide nanowires and axial hetero-structure nanowire arrays using the technique of glancing angle deposition and discusses many related applications such as photodetector and wettability applications. It is shown that the glancing angle deposition technique is simple, cost-effective, and catalytic free with many advantages compared to other methods.

Section 2 mainly focuses on group III–V compounds, including GaN for photoelectrochemical water splitting, InGaN nanowires for photovoltaic applications, and in situ techniques facilitating understanding of GaAs nanowire growth. Si is one of the most known bulk semiconductors and the logic device basis-CMOS structure is largely based on Si material. However, group III–V semiconductors, such as GaN, GaAs, and InN, provide numerous property advantages over Si. They demonstrate outstanding electronic and optical properties, for instance, direct band gap, increased carrier mobility, and low exciton binding energy, which give these semiconductors great potential for use in optoelectronic and microelectronic fields. Chapter 4 shows that GaN is a promising photoelectrode for photoelectrochemical water spitting reaction due to its tunable band gap, favorable band edge positions, and extraordinary stability. Chapter 5 presents a comparative analysis of different structural formats of InGaN, such as planar, nanodisk, and core-shell-shell nanowires, for their performances as potential photovoltaic material and concludes that nanowire-type structure displays a better performance. Chapter 6 introduces advanced in situ techniques to provide direct interpretation and time-resolved observation of the growth mechanism of nanowires, which allows better control of nanowire growth for specific applications.

Section 3 contributes to the fundamental understanding of other nanowire structures including basic Si nanowires, metallic wires, and complex geometrical nanowires. Chapter 7 reports a growth method of indium-catalyzed Si nanowires via vapor-liquid-solid mode for nanoscale device applications, with good control of scaling down the diameter of the Si nanowire and an enhanced growth density of nanowires. In addition to the semiconductor nanowires already mentioned, Chapter 8 focuses on a study of a hybridization design between plasmonic and photonic-guided modes in periodic arrays of metallic nanowires integrated on top of dielectric waveguides. The chapter aims to stimulate further research efforts in the development of integrated hybrid photonic–plasmonic devices in the research community. The last two chapters introduce complex geometrical nanowires. Chapter 9 provides a thorough study on three-nanometer node gate-all-around field-effect transistors based on two types of nanostructures: nanowire and nanosheet. The study suggests that nanowire field-effect transistors have better performance compared to the fin-shaped and nanosheet transistors at low power applications. Chapter 10 suggests a method to engineer an enhanced donor-acceptor behavior in blending and core-shell nanowires. It describes a synthesis method of poly (para-phenylene-vinylene) PPV (electron donor) and poly (vinyl-carbazole) PVK (good hole transport) nanowires, and its core-shell architecture bases on PPV and PVK polymers. This study found that the core-shell morphology based on PVK and PPV polymers showed an amplified emission of PPV intensity by adding PVK. The chapter provides a way to develop alternative in-solution processing techniques to manage the local organization of donor-acceptor systems at the scale of the exciton diffusion length.

Finally, I would like to acknowledge all the contributors for their hard work. I am also grateful for the opportunity provided by IntechOpen, such that this interesting topic can be reviewed and shared among the research community. I would also like to express my appreciation for all the excellent editing work done by the IntechOpen staff. At last, my special acknowledgment goes to Author Service Manager Josip Knapić for all of his encouragement, patience, and kindness during the editing process.

Dr. Xihong Peng
College of Integrative Sciences and Arts,
Arizona State University,
Polytechnic Campus,
Arizona, United States

Section 1

Oxide Nanowires

Chapter 1

ZnO Nanowire Field-Effect Transistor for Biosensing: A Review

Nonofo Mathiba Jack Ditshego

Abstract

The last 19 years have seen intense research made on zinc oxide (ZnO) material, mainly due to the ability of converting the natural n-type material into p-type. For a long time, the p-type state was impossible to attain and maintain. This chapter focuses on ways of improving the doped ZnO material which acts as a channel for nanowire field-effect transistor (NWFET) and biosensor. The biosensor has specific binding which is called functionalization that is achieved by attaching a variety of compounds on the designated sensing area. Reference electrodes and buffers are used as controllers. Top-down fabrication processes are preferred over bottom-up because they pave way for mass production. Different growth techniques are reviewed and discussed. Strengths and weaknesses of the FET and sensor are also reviewed.

Keywords: zinc oxide (ZnO), semiconductor device, nanosensor, nanowire field-effect transistor (NWFET), biosensors, growth techniques

1. Introduction

Zinc oxide (ZnO) material has been known as a semiconductor for over 70 years, with some of the first literature being reported as early as in 1944 [1]. It was never put to use like other semiconductors (GaN, Si) because it is difficult to dope. The past 19 years have seen a revival on the research and use of material because of new and emerging ways of doping it. The material is naturally n-type [1–4], and by controlling the conditions of growth, the donor concentration can be controlled. The growth conditions include temperature, diethyl zinc (DEZ) reactant, O_2 or H_2O reactant, and pressure. P-type material [1–4] is difficult to grow and tends to slowly revert back to n-type. Researchers [5–14] who managed to deposit the p-type material have shown that it converts back to n-type within a few days. Maximum time period shown on p-type ZnO was a few months [5–14].

ZnO is a wide bandgap semiconductor [e.g., (0 K) = (3.441 ± 0.003) eV; (300 K) = (3.365 ± 0.005) eV]. It belongs to the group of II^b-VI compound semiconductors which crystalize exclusively in the hexagonal wurtzite-type structure. The lattice parameters of the wurtzite crystal structure are: a = 3.24 Å and c = 5.21 Å. Related to similar II^b-VI (e.g., CbS, CbSe, ZnSe, and ZnS) or III-V (e.g., AlSb, Bas, GaN, and InSb) semiconductors, it has comparatively strong polar binding and large exciton binding energy of (59.5 ± 0.5) meV. Its density is 5.6 g cm^{-3}, a value which corresponds to 4.2×10^{22} ZnO molecules per cm^{-3} [1, 2].

ZnO has practical advantages that make it an attractive semiconductor from an industrial point of view. It has low cost; is abundant, nontoxic, and transparent; has large excitonic binding energy of 60 meV; is soluble, compatible with intercellular material; and has wide and direct bandgap of 3.37 eV, making it highly sensitive. It is well known that semiconductors have a small bandgap which allows switching between conduction and off-states. The larger the bandgap, the better the semiconductor is able to switch states and insulate leakage currents. Bandgap affects sensitivity because a device that possesses a wider bandgap allows for higher currents to travel but also prevents leakage currents, which results in more sensitive and accurate readings. With low-temperature fabrication processes, high-quality devices can be fabricated using the conventional processing technology, thereby making it suitable for low-cost mass-production. It has potential applications in optoelectronics, transparent electronics, and spintronics. ZnO and its alloys have versatile electrical and optical properties for applications in thin film or nanowire transistors, light emitters, biosensors, and solar cells. The nanowire biosensor has a high surface-to-volume ratio, enabling real-time and label-free detection [1–4, 15–17].

Currently, the main commercial application for ZnO (and/or IGZO) material is in displays, with companies like Sharp and Samsung putting IGZO into mobile phone displays [18–20]. IGZO displays outperform other semiconductor displays such as amorphous silicon and organic semiconductors by providing improved resolution and reduced power consumption. This is possible because IGZO has a 20× to 50× times higher mobility than amorphous silicon and polymers, which allows for device scaling without affecting performance [18–20]. Higher mobility values can also be achieved with amorphous silicon technology, but it needs to be laser annealed which is expensive.

2. Growth techniques of ZnO

ZnO films can be grown using three methods: gas transport (vapor phase deposition), hydrothermal synthesis, and/or melt process. Melt growth techniques are a problem due to high vapor pressure of ZnO. Growth using gas transport is difficult to control for large film layers and is normally used for bottom-up ZnO nanostructures. Hydrothermal synthesis is therefore preferred as a method of growth. Thin films can be produced through chemical vapor deposition, metalorganic vapor phase epitaxy, electrodeposition, pulsed laser deposition, sputtering, sol–gel synthesis, atomic layer deposition, spray pyrolysis, etc. All the mentioned techniques fall under hydrothermal synthesis, and one of the preferred methods is atomic layer deposition (ALD). The ALD process is capable of producing highly conformal and quality films [21]. The process is cyclic and is based on the number of reactants. **Figure 1** shows that the ALD process for ZnO films is cyclic and depends on two reactants: metallization and oxidation.

Metallization uses diethyl zinc (DEZ) as the zinc (Zn) metal precursor. Purge and pump steps are used to separate the execution of the reactants and to remove by-products. Before deposition, the wafer (substrate) is preheated at a temperature that will be used for deposition and it is also cleaned with O_2 plasma so as to remove any polymer layer. During the metallization step, the DEZ (Zn $(C_2H_5)_2$) is absorbed onto the surface of the wafer and the residual Zn $(C_2H_5)_2$ is removed from chamber. "R" in **Figure 1** represents C_2H_5.Then on another step, water or O_2 is delivered to react with the absorbed DEZ [23–25]. These steps are executed separately, and to ensure this, purge steps are introduced in between the steps.

When water is used instead of O_2 for oxidation, the process is called thermal ALD. This process tends to produce films similar to chemical vapor deposition (CVD)

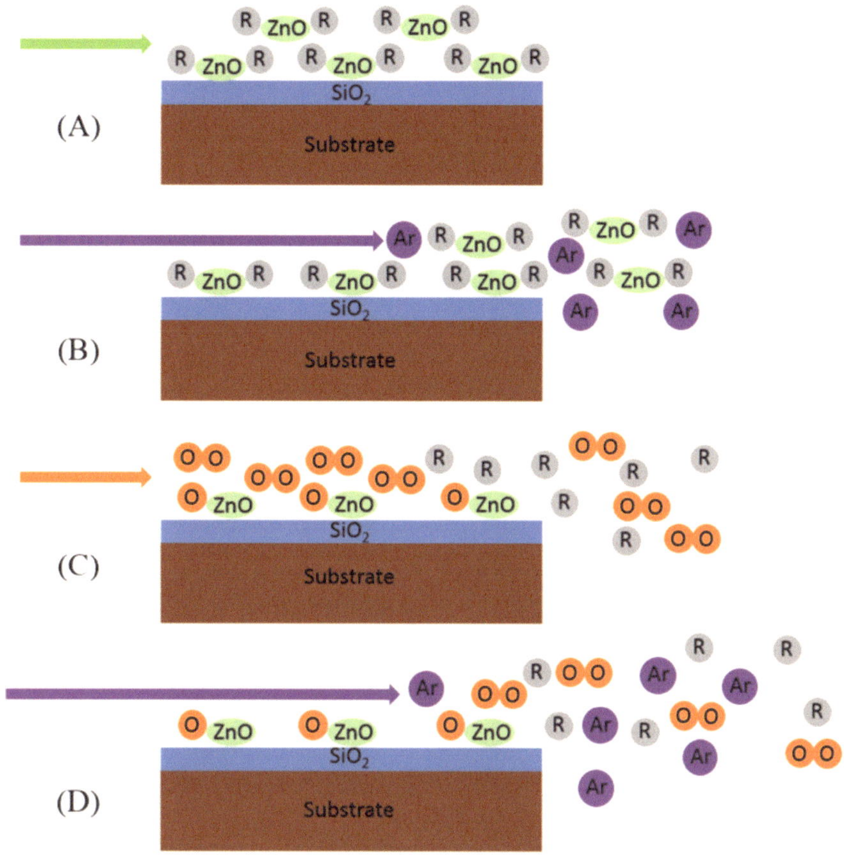

Figure 1.
Schematic diagram illustrating a single cycle of ZnO deposition using the ALD tool (A) metallisation and oxidation step, (B) Purge and pump step (C) Cleaning with O2 plasma step, (D) Removing non-used products with Ar step [22].

techniques [25–27]. When O_2 is used instead of water, then the process needs plasma energy. Remote plasma atomic layer deposition (RPALD) is a fairly new process which is why it is still not in used. It is better than the other deposition techniques as it tends to produce films close to epitaxial layers. The layers are crystalline but tend to be nonuniform to the underlining layer which is why they are not called epitaxial layers. It is a process with great potential for depositing highly conformal and quality films. The process is better than thermal ALD in terms of conformity and quality, but both processes do not generally produce epitaxial layers due to nonuniformity to the underlining substrate. The plasma-assisted ALD method has the following advantages: reduction of OH impurity, allows more freedom in processing conditions, and provides wider range of material properties. The OH impurity is not desired as it affects the conductivity of the semiconductor and induces defects in the dielectrics.

Table 1 compares various growth techniques and how they affect NWFET output characteristics. Chemical vapor deposition (CVD) is the most popular technique for bottom-up nanowire processes. There are two growth techniques classified under CVD which are vapor–liquid–solid (VLS) and vapor–solid (VS) deposition techniques. CVD normally give the highest mobility as they produce crystalline wires with the only flaw being from the catalysts that guide the growth. VS produces better quality nanowires than VLS as it uses no catalysts but instead uses very high temperatures (>900°C). The problem with VS is that it is usually harder to control the size and morphology of the nanowires.

No	Processing route	Synthesis method	Starting materials	Synthesis temp. (°C)	Morphology	Diameter of ZnO nanostructure (mm)	Length of ZnO nanostructure	Ref.
1	Vapor phase processing	Thermal evaporation	Zn metal, O_2, and Ar	650–670	Nanowire	100	Several microns	[29]
2	Route		Zn metal pellets, O_2, Ar	900	Nanowire	20	—	[30]
3			Zn powder, O_2, Ar	600	Nanowire	80	1 μm	[31]
4		Vapor phase transport	ZnO powder, graphite, Cu catalyst	930	Hierarchical dendrite	60–800	—	[32]
5		Aerosol	Zn powder, N_2 gas	500–750	Fiber-mat	100–300	—	[33]
6					Cauliflower	20–30	—	
7		RF sputtering	ZnO deposited over Pt sputtered interdigitated alumina substrate	—	Nanobelt	—	Few micrometer	[34]
8		Molecular beam epitaxy	Zn metal, O_3/O_2 plasma discharge, Au coated substrate	600	Nanorod	50–150	2–10 μm	[35]
9	Solid-state processing	Carbothermal reduction	ZnO powder, graphite powder, Ar gas flow, Au coated silicon substrate	900–925	Nanowire	80–120	10–20 μm	[36, 37]
10	Route	Solid-state Chemical	$ZnCl_2$, NaOH, polyethylene Glycol, $Na_2WO_4.2H_2O$	RT	Nanorod	40–60	200 nm	[38]
11		Reaction				20–40	100 nm	
12	Wet processing	Hydrothermal	$ZnAc^2$, NaOH, absolute ethanol, distilled water	180	Nanorod	—	—	[39]
13	Route		$Zn(CH_3COO)_2{\cdot}2H_2O$, $C_6H_8O_7{\cdot}H_2O$, absolute ethanol, distilled water	400	Nanorod (vertically aligned)	50	500 nm	[40]
14			$Zn(NO_3)_2{\cdot}6H_2O$, NaOH, cetyltrimethyl ammonium bromide, ethanol	120	Nanorod	—	—	[41]

DOI: http://dx.doi.org/10.5772/intechopen.93707

No	Processing route	Synthesis method	Starting materials	Synthesis temp. (°C)	Morphology	Diameter of ZnO nanostructure (mm)	Length of ZnO nanostructure	Ref.
15			$Zn(NO_3)_2{\cdot}6H_2O$, NaOH, cyclohexylamine, ethanol, water	200	Nanorod	150–200	2 μm	[42]
16			$Zn(SO_4){\cdot}7H_2O$, NH_4OH, deionized water	75–95	Nanorod	—	—	[43]
17		ALD	DEZ ($Zn(C_2H_5)_2$), H_2O	—	Nanowire	70–100	5 μm	[44]
18		Plasma ALD	DEZ ($Zn(C_2H_5)_2$), O_2	150–190	Nanowire	36–100	2–20 μm	[22]

Table 1.
Summary of various methods used for the production of 1-D ZnO nanostructures, adopted from [28].

Table 1 also shows that atomic layer deposition (ALD) is an attractive technique because it deposits high quality films at low temperatures between 120 and 210°C [22, 45]. The problem with ALD is that it has only this window for good quality conducting films. At temperatures below 120°C, the deposition can be incomplete or experience condensation depending on growth rate. At temperatures above 210°C, the deposition tends to experience desorption or it decomposes toward CVD deposition. Nonetheless, it is one of the best techniques toward growing films close to epitaxial growth (crystallinity is achievable whereas uniformity is still difficult to achieve) [22, 45]. The tool has shown potential by achieving high values of field effect mobility >30 cm^2/Vs with excellent crystallinity.

2.1 Native point defects

There are three types of defects in a crystal lattice: point defects, area defects, and volume defects. Point defects which are caused by native elements and impurities are the major problem for ZnO semiconductor. Native point defects for ZnO include the following: zinc interstitial (Zn_i), zinc antisite (Zn_o), zinc vacancy (V_{Zn}), oxygen interstitial (O_i), oxygen antisite (O_{Zn}), and oxygen vacancy (V_o). Over the years, a lot of research advocated them as the major cause for the n-type behavior. Oxygen defects are seen as the main contributors toward the n-type behavior [3, 15]. There are some researchers [1–4] who hypothesize that impurities (not the native point defects) are the main cause of the n-type behavior because they tend to be shallow donors whereas Zn and O_2 defects tend to be deep donors [1–4]. The two theories have not been proven so currently the main cause of the natural n-type behavior of ZnO [1–4] is not certain.

2.2 Deep donors versus shallow donors: ZnO

ZnO impurities (foreign atoms) are normally incorporated in the crystal structure of the semiconductor. There are two reasons of impurity incorporation: they can either be unintentionally introduced due to lack of control during growth processes or they are intentionally added to increase the number of free carriers in the semiconductor. Impurities in the ZnO should have the ability to be ionized; which is desirable as it increases conductivity. This means that the impurity atoms should be able to give off electrons to the conduction band. If the impurities were acceptors—they should be able to give off holes to the valence band [3, 16].

Donor Impurities for the n-type ZnO can either be shallow or deep. **Figure 2** shows shallow donors compared to deep donors. Shallow impurities require little energy to ionize (this is energy typically around the thermal energy or less). These donor impurities possess energy close to the band edge—the extra valence electron

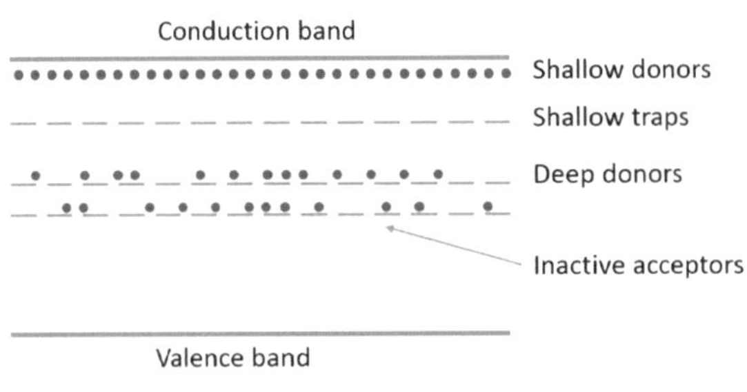

Figure 2.
Shallow versus deep donors [1–4].

of these impurities are loosely bound and occupy effective-mass states near the conduction band maximum- CBM- at low temperatures. Deep impurities on-the-other-hand require energy greater that the thermal energy to ionize. These donor impurities possess energy far from the band edge (CBM) making them very hard to ionize. Their presence within the semiconductor tends to contribute only a small fraction of free carriers. Deep donors are also called traps because they act as effective recombination centers in which electrons and holes fall and annihilate each other. Grain boundaries (GB) are main source of deep state impurities and they adversely affect transistor performance. ZnO is a wide bandgap material and research suggests [3, 4, 16] that there exist possible deep-level traps in GBs. The examples of deep donors are Zn and O ions. Zn acts as a deep donor when there is a vacancy and O acts as a deep donor in any defect state. An example of a shallow donor is the H ion.

2.3 Top-down fabrication of ZnO nanowire FETs

There are four main methods capable of producing nanometer features using top-down approaches: UV stepper lithography, e-beam lithography [46], focused ion-beam lithography [47], and spacer method [45, 48]. UV lithography is the standard industrial method for fabricating nanodevices. E-beam and focused ion-beam lithography are often used and can pattern devices down to 5 nm, but the equipment is very expensive and the pattern writing is very slow. These two instruments resemble scanning electron microscope (SEM) in terms of operation. Whereas SEM is used to focus a beam of electrons to image samples within a chamber, these instruments are used to create patterns on the samples. The difference between e-beam and focused ion-beam is that the latter uses an ion beam to pattern wafers and hence does not require photoresist. Their advantage over optical UV lithography is the small features they reach. For low-cost applications such as biosensors, the problem with these two methods is that they are expensive.

The spacer technique is a low-cost fabrication method for fabricating nanowires. It was first reported in 2005 by Ge et al. [49], and other researchers [44, 50, 51] have since carried it forward. The technique has great potential in shaping nanometer features using conventional, low-cost photolithography. **Figure 3** shows the concept of the spacer technique. It uses first anisotropic etch to create a vertical pillar on an insulating layer (SiO_2), then after deposition of a semiconductor layer (ZnO) and a second anisotropic etch, to create nanowires made up of the semiconductor layer. This method allows nanowire features with controllable dimensions to be developed. The ICP tool is usually used for anisotropic etching and produces

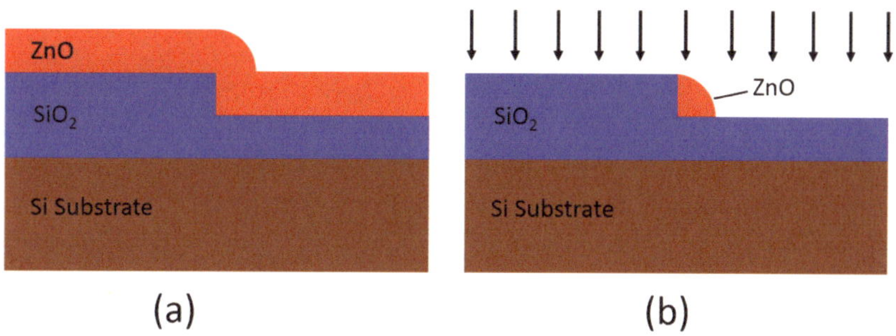

Figure 3.
Novel spacer technique used to pattern nanowire features. Cross-sectional schematic of nanowire formation (a) before dry etch and (b) after dry etch [22].

surface roughness <1.5 nm. Other tools such as RIE and ion beam etch produce roughness >5 nm. The fabrication process for the complete ZnO NWFET structure is as outlined in [52].

3. Background on FETs

The ZnO field-effect transistor (FET) has been around for decades. The success of the device in meeting the technological demands has largely been dominated by the shrinking size of its physical geometry. It has an advantage as a junctionless (no p-n junctions) FET compared to conventional FETs [17, 21, 23–27, 53, 54]. There has been an introduction of new materials and heterojunction structures developed so as to move away from conventional silicon devices. High-K dielectrics have been introduced to replace the conventional SiO_2 which should help maintain acceptable dielectric thicknesses while keeping gate leakage currents low [17, 21, 23–27, 53, 54].

Even with so many improvements being made to the device, the limits of FET scaling are approaching. The thickness of the oxide (t_{ox}) cannot be less than 1 nm due to high tunneling current and significant operational variation. The substrate doping is also very high which creates leakage and tunneling currents that are unacceptable to device operation.

3.1 ZnO thin film transistors (TFTs)

TFTs have also been fabricated using ZnO, mainly as thin film transistors for application in displays. **Figure 4** compares 20 ZnO TFTs fabricated by different authors [27, 53–71] using a variety of fabrication methods over the last 5 years. The graph is a plot of field effect mobility versus subthreshold slope which are two of the main parameters that describe the performance and efficiency of a device. The best device was fabricated by Bayraktaroglu et al. [70] with a SiO_2 insulator and pulsed laser-deposited ZnO active channel layer. The device had a field effect mobility 110 cm^2/Vs and an excellent subthreshold gate voltage swing of 109 mV/decade. This value of mobility is much higher than the value of around 1 cm^2/Vs that is typically achieved with amorphous silicon TFTs in production displays. It is clear therefore that ZnO TFTs have considerable potential for application in high performance displays.

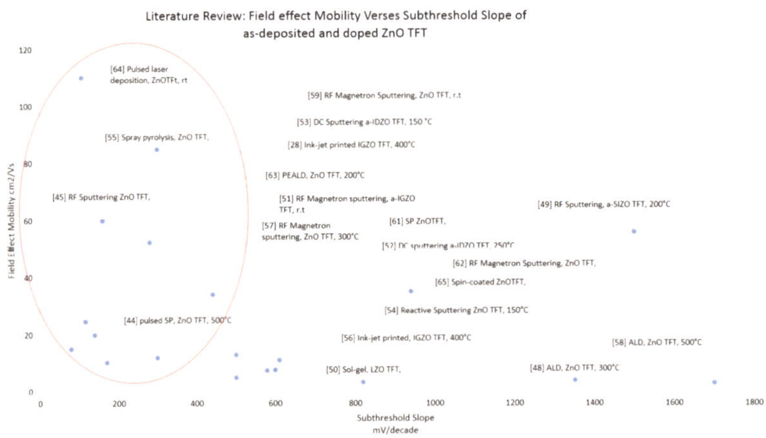

Figure 4.
General literature review on TFTs looking at field effect mobility versus subthreshold slope of as-deposited and doped ZnO films.

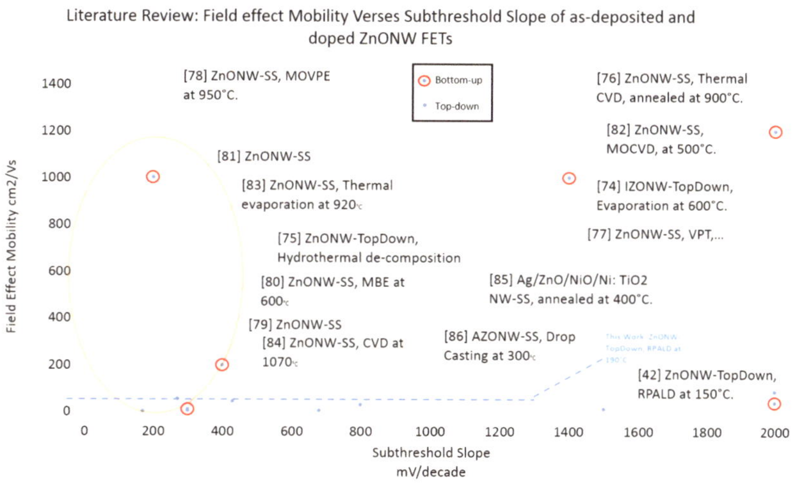

Figure 5.
Literature review on nanowire FETs looking at field effect mobility versus subthreshold slope of as-deposited and doped ZnO nanowires.

3.2 Nanowire field-effect transistors (FETs)

Emerging nonplanar devices [17, 21] are being researched to prolong the future progress for FETs. Devices based on quasi-one-dimensional (1-D) nanostructures are still at an embryonic stage from an industrial point of view. These nanostructures include the following: nanowires, nanobelts, nanoribbons, and nanoneedles [72, 73]. This review is interested in nanowire FETs which are also being researched for application in biosensors because the high surface-to-volume ratio provides high sensitivity.

3.3 Comparing ZnO NWFETs

Figure 5 compares 15 different ZnO NWFETs fabricated by different authors using a variety of methods [22, 74–86]. The graph is plotted with field effect mobility against the subthreshold slope, which are two important device parameters that determine ZnO NWFET performance. The nanowires were fabricated using top-down and bottom-up (self-assembled) processes. Self-assembled processes tend to display very high field effect mobility which is normally above 200 cm^2/Vs; whereas the top-down have lower mobility values. Most of the top-down fabricated devices have mobility <1.0 cm^2/Vs with around three papers giving a mobility >10.0 cm^2/Vs. The difference in the mobility may be due to the fact that self-assembled nanowires are single-crystal, whereas top-down nanowires are polycrystalline. Nonetheless, top-down techniques are desirable as they currently pave way for mass production and will be pursued in this research investigation.

4. Biosensors

A biosensor is defined by the International Union of Pure and Applied Chemistry (IUPAC) as "a self-contained integrated device that is capable of providing specific quantitative or semiquantitative analytical information using a biological recognition element (biochemical receptor), which is retained in contact direct with a transduction element" [87]. A biosensor is a "more-than-Moore device" because it incorporates functionalities that do not necessarily scale according to Moore's law. Under the roadmap, the device falls under the category of sensors and actuators. Other categories include analogue/RF, passives, HV power, and biochips [88, 89].

Figure 6 shows a typical structure of a biosensor [90–92]. The biomolecules are contained within an analytic solution and attach themselves to immobilized enzymes or immune-agents on the linkers. Linkers in turn are attached to the transducer. The transducer then converts the charge on the analyte into an electrical signal which is then transmitted for data processing. Biosensors can be considered as part of the research field known as "chemical sensors" in that a biological mechanism is used for analyte detection within an analyte solution [93–95]. Quasi-one-dimensional nanostructures have a greater surface-to-volume ratio compared to planar structures and are therefore expected to be more sensitive than planar sensors [93–95].

Nanowires are the same as nanorods. The words can be used interchangeably [80]. These have received enormous attention due to their suitable properties for designing novel nanoscale biosensors. For example, the dimensions of ~1–100 nm are similar to those of many biological entities, such as nucleic acids, proteins, viruses, and cells [79]. In addition, the high surface-to-volume ratios for nanomaterials allow a large proportion of atoms in the bio-analyte to be located at or close to the surface. Moreover, some nanowire materials have surfaces that can easily be chemically

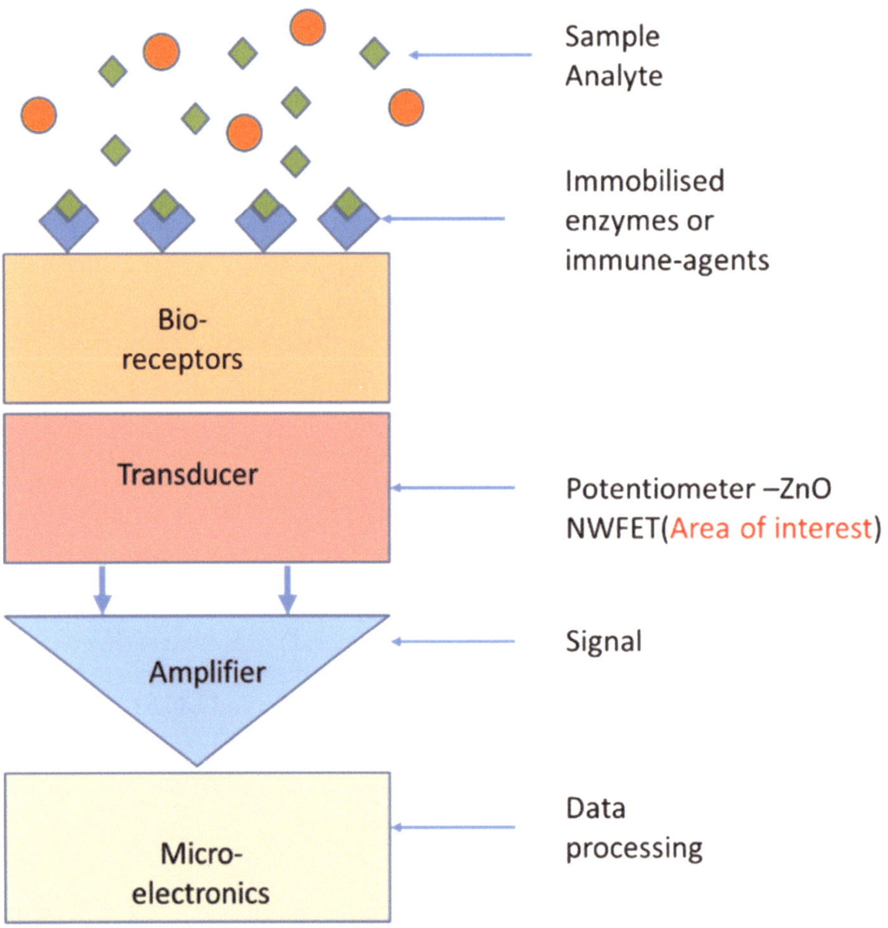

Figure 6.
Typical structure of a biosensor. The biomolecules are contained within an analytic solution and attach themselves to immobilized enzymes or immune-agents on the receptors. The transducer then converts the energy signal produced into an electrical signal which is then transmitted for data processing. [22].

modified which makes them significant candidates for biosensors [79, 80]. There are a number of nanostructure-based electrical biosensors which include single-wall carbon nanotubes (SWCNT), nanowires, nanogaps, nanochannels, and nano-electromechanical (NEM) devices. The project will focus on nanowire-based devices as they have considerable potential for electrical biosensing that offer the possibility of portable assays in a variety of point-of-care environments [48, 90, 96].

4.1 Silicon biosensors

Over the past decade, silicon nanowires have been the most researched for application as biochemical sensors [97–108]. Silicon nanowires are of interest for a number of reasons, for example, the material is well known and is compatible with CMOS integrated circuits for the development of sensor systems [97–108]. The nanowire is expected to have high surface-to-volume ratios which give high sensitivity and the electrical sensing will give real-time label-free detection without the use of expensive optical components. Mass manufacturing is also a main advantage for silicon and is critically important for nanowire biosensor applications because of the widespread uptake of biosensors in "point-of-care" settings, the biosensor needs to be disposable [97–108].

A number of fabrication methods are well established for silicon nanowires which utilize both bottom-up and top-down methods (these methods are called hybrids). It still remains that bottom-up techniques have the advantage of simplicity [97–108]. Bottom-up methods are still limited due to the alignment problem. The hybrid methods require further nanowire technologies to achieve alignment, such as electric field or fluid-flow-assisted nanowire positioning to locate the nanowires between lithographically defined source and drain electrodes. The technique is interpreted as a hybrid between bottom-up and top-down. Top-down methods overcome these problems, and several researchers have used advanced lithography techniques to fabricate single-crystal silicon nanowires on silicon-on-insulator (SOI) substrates. SOI wafers are expensive and to overcome the problem some researchers [109] have devised alternatives to SOI. The electrical output characteristics of silicon nanowires are good and they are well suited for biosensing applications. The sensitivity range for most silicon-nanowire based biosensors is between 50 and 400 mV [97–134].

4.2 Comparing ZnO nanowire biosensors

ZnO is investigated as it is expected to be more sensitive than Si due to its wider bandgap [109]. This is observed by comparing **Table 2** with **Table 3**. ZnO devices show results comparable to silicon devices; especially looking at response time and limit of detection. It is required that biosensors should have the liquid reference electrode. There are many different types of ZnO nanostructures being used for sensing application and **Table 2** compares the ZnO nanostructures such as nanotetrapods, nanocombs, and nanorods used for biosensing [110, 121]. Nanotetrapods [123] are like nanorods but with four single crystalline legs. Most of the ZnO devices were synthesized by vapor phase method and then transferred on Au electrode to form a multiterminal network for the sensor receptors. Like all other bottom-up ZnO nanostructures discussed here, they are transferred to a surface of a working electrode to form a thin layer to modify the transducer. The devices have low sensitivity but the nanotetrapods exhibit good detection limit down to ~1.0 nM. The researchers [123] did not explain why the nanostructures possess low sensitivity but its three-dimensional features have the potential for multiterminal communication applications [123].

No.	Reference electrode	Type of sensor	Channel material	ZnO fabrication process	LOD (μM)	Response time (s)	Ref.
1	Au	Biosensor	ZnO nanorod array	Hydrothermal	10	<5	[111]
2	ITO	Biosensor	ZnO nanotube array	Hydrothermal/ chemical	10	<6	[112]
3	Au	Biosensor	Tetrapod-like ZnO	CVD	4	6	[113]
4	Glass capillary	Biosensor	ZnO nanoflakes	Hydrothermal	0.5	<4	[114]
5	GCE	Biosensor	Fork-like ZnO	Annealing	0.3	3	[115]
6	Au	Biosensor	Comb-like ZnO	CVD	20	<10	[116]
7	Ti	Biosensor	ZnO/C nanorod array	Hydrothermal	1	4	[117]
8	ITO	Biosensor	ZnO/ Cu array matrix	Hydrothermal	40	<6	[118]
9	GCE	Biosensor	ZnO/Au nanorods	Hydrothermal	0.01	<5	[119]
10	Pt	Biosensor	ZnO/NiO nanorods	Hydrothermal	2.5	<5	[120]

Table 2.
Summary of characteristics for various 1-D ZnO biosensors, adopted from [110].

No.	Reference electrode	Type of sensor	Channel material	ZnO fabrication process	LOD (pM)	Response time (s)	Ref.
1	No reference electrode	Biosensor	Si NW	nanocluster-mediated vapor–liquid–solid growth method	10	<10	[97]
2	Au	Biosensor	Si NW	Chemical vapor deposition	0.002	<10	[98]
3	Platinum wire	Biosensor	Si NW	SNAP technique	10	<10	[101]
4	None	Biosensor	Si NW	Reactive-ion etching (RIE)	0.01	<10	[106]
5	None	Biosensor	Si NW	Synthesized by chemical vapor deposition	100	<10	[122]

Table 3.
Summary of characteristics for various 1-D Si biosensors, adopted from [121].

In nanocombs [116] design, each comb has between 3 and 10 rods connected to one another by a single rod. ZnO nanocombs were used as the channel for sensing glucose [116] and as label-free uric acid biosensor based on uricase [124]. The functionalized

ZnO nanorods showed thermal stability, anti-interference capability, and direct electron transfer (DET) between enzyme electroactive sites and external electrodes. The activity of the enzyme and the sensitivity can be increased by introducing a lipid film between the channel and the enzyme. Another uric acid biosensor [125] example is based on uricase-functionalized ZnO nanoflakes, which was hydrothermally prepared at low temperatures on Au-coated glass. The sensor produced a sensitivity based on subthreshold slope of ~66 mV/decade. Bottom-up ZnO nanorods [126] were also used as lactate oxidase (LOD) biosensor using glutaraldehyde cross-linkers. The device had a subthreshold sensitivity of ~41 mV/decade, with maximum detection of 0.1 μM. To test for cholesterol, porous ZnO mirco-tubes [127] were constructed using 3-D assembled porous flakes. ZnO nanorods [128] were grown on Ag electrode to make a cholesterol sensor.

5. Conclusion

Most researchers use bottom-up approaches to fabricate the ZnO biosensors because of the straightforward synthesis process. However, these bottom-up devices have variable electrical performance due to the lack of geometrical dimension control and addressing the nanostructures for sensing application. So far, there is limited research reported on top-down ZnO biosensors, and previous work demonstrated the viability of top-down ZnO NWFET for biosensor applications. In the work, however, there was no passivation layer on the ZnO nanowires, which led to the dissolution of the material. This made the device unstable and the sensing results were not reproducible. There exists a need to develop a passivating layer technology and optimize the fabrication process for biosensor applications. That way, a reliable measurement of sensitivity for the nonspecific and specific sensing of lysozyme and bovine serum albumin (BSA) can be achieved.

Acknowledgements

N.M.J. Ditshego would like to acknowledge the Botswana International University of Science and Technology (BIUST) for supporting his doctoral studies and the Southampton Nanofabrication Centre for the experimental work. The author would like to acknowledge the EPSRC EP/K502327/1 grant support.

Author details

Nonofo Mathiba Jack Ditshego
Electrical, Computer and Telecommunications Engineering Department, CET, Botswana International University of Science and Technology (BIUST), Palapye, Botswana

*Address all correspondence to: ditshegon@biust.ac.bw

References

[1] Klingshirn C, Fallert J, Zhou H, Sartor J, Thiele C, Maier-Flaig F, et al. 65 years of ZnO research—old and very recent results. Physica Status Solidi. 2010;**247**(6):1424-1447

[2] Özgür U, Alivov YI, Liu C, Teke A, Reshchikov MA, Doğan S, et al. A comprehensive review of ZnO materials and devices. Journal of Applied Physics. 2005;**98**(4):041301

[3] McCluskey MD, Jokela SJ. Defects in ZnO. Journal of Applied Physics. 2009;**106**(7):071101

[4] Janotti A, Van de Walle CG. Fundamentals of zinc oxide as a semiconductor. Reports on Progress in Physics. 2009;**72**(12):126501

[5] Wang G, Chu S, Zhan N, Zhou H, Liu J. Synthesis and characterization of Ag-doped p-type ZnO nanowires. Applied Physics A: Materials Science & Processing. 2011;**103**(4):951-954

[6] Tsukazaki A, Ohtomo A, Onuma T, Ohtani M, Makino T, Sumiya M, et al. Repeated temperature modulation epitaxy for p-type doping and light-emitting diode based on ZnO. Nature Materials. 2005;**4**(1):42-46

[7] Oh MS, Hwang DK, Lee K, Im S, Yi S. Low voltage complementary thin-film transistor inverters with pentacene-ZnO hybrid channels on Al Ox dielectric. Applied Physics Letters. 2007;**90**(17):173511-173513

[8] Choi J-H, Lee SW, Kar JP, Das SN, Jeon J, Moon K-J, et al. Random network transistor arrays of embedded ZnO nanorods in ion-gel gate dielectric. Journal of Materials Chemistry. 2010;**20**(35):7393

[9] Dunlop L, Kursumovic A, MacManus-Driscoll JL. Reproducible growth of p-type ZnO:N using a modified atomic layer deposition process combined with dark annealing. Applied Physics Letters. 2008;**93**(17):172111

[10] Huang J, Chu S, Kong J, Zhang L, Schwarz CM, Wang G, et al. ZnO p-n homojunction random laser diode based on nitrogen-doped nanowires. Advanced Optical Materials. 2013;**1**(2):179-185

[11] Kumar M, Kar JP, Kim IS, Choi SY, Myoung JM. Fabrication of As-doped p-type ZnO thin film and ZnO nanowire inserted p-n homojunction structure. Applied Physics A: Materials Science & Processing. 2009;**97**(3):689-692

[12] Yuan GD, Zhang WJ, Jie JS, Fan X, Zapien JA, Leung YH, et al. p-type ZnO nanowire arrays. Nano Letters. 2008;**8**(8):2591-2597

[13] Cha SN, Jang JE, Choi Y, Amaratunga GAJ, Ho GW, Welland ME, et al. High performance ZnO nanowire field effect transistor using self-aligned nanogap gate electrodes. Applied Physics Letters. 2006;**89**(26)

[14] Li FM, Hsieh G-W, Dalal S, Newton MC, Stott JE, Hiralal P, et al. Zinc oxide nanostructures and high electron mobility Nanocomposite thin film transistors. IEEE Transactions on Electron Devices. 2008;**55**(11):3001-3011

[15] Murphy TE, Blaszczak JO, Moazzami K, Bowen WE, Phillips JD. Properties of electrical contacts on bulk and epitaxial n-type ZnO. Journal of Electronic Materials. 2005;**34**(4):389-394

[16] Lee D-J, Kwon J-Y, Kim S-H, Kim H-M, Kim K-B. Effect of Al distribution on carrier generation of atomic layer

deposited Al-doped ZnO films. Journal of the Electrochemical Society. 2011;**158**(5):D277

[17] ZnO Devices and Applications: A Review of Current Status and Future Prospects, IEEE Xplore, Proceedings of the IEEE. August 2010;**98**(7):1255-1268. DOI: 10.1109/JPROC.2010.2044550

[18] Triggs R. Display Technology Explained: A-Si, LTPS, Amorphous IGZO, and Beyond (Android Authority). 2014. Available from: http://www.androidauthority.com/amorphous-igzo-and-beyond-399778/ [Accessed: 01 March 2015]

[19] Hruska J. The Perils and Promise of High-Resolution Displays (Extreme Tech) [Online]. 2012. Available from: http://www.extremetech.com/electronics/126519-the-perils-and-promise-of-high-resolution-displays [Accessed: 01 March 2015]

[20] Anthony S. IGZO Display Tech Finally makes it to Mass Market: iPad Air Now, High-res Desktop Display Soon (Extreme Tech) [Online]. 2013. Available from: http://www.extremetech.com/computing/170970-igzo-display-tech-finally-makes-it-to-mass-market-ipad-air-now-high-res-laptops-and-desktops-next [Accessed: 01 March 2015]

[21] Riyadi MA, Suseno JE, Ismail R. The future of non-planar nanoelectronics MOSFET devices: A review. Journal of Applied Sciences. 2010;**10**(18):2136-2146

[22] Sultan SSM. Top-down fabrication and characterization of zinc oxide nanowire field effect transistors [PhD thesis]. University of Southampton; 2013

[23] Park J-S. The annealing effect on properties of ZnO thin film transistors with Ti/Pt source-drain contact. Journal of Electroceramics. 2010;**25**(2-4):145-149

[24] Engineering E. Amorphous In-Ga-Zn-O Thin Film Transistor for Future Optoelectronics by Tze-Ching Fung. 2010

[25] Moon Y, Lee S, Kim D. Characteristics of ZnO based TFT using La2O3 high-k dielectrics. Journal of Korean. 2009;**55**(5):1906-1909

[26] Park SY, Kim BJ, Kim K, Kang MS, Lim KH, Il Lee T, et al. Low-temperature, solution-processed and alkali metal doped ZnO for high-performance thin-film transistors. Advanced Materials. 2012;**24**(6):834-838

[27] Ortel M, Wagner V. Leidenfrost temperature related CVD-like growth mechanism in ZnO-TFTs deposited by pulsed spray pyrolysis. Journal of Crystal Growth. 2013;**363**:185-189

[28] Arafat MM, Dinan B, Akbar SA, Haseeb ASMA. Gas sensors based on one dimensional nanostructured metal-oxides: A review. Sensors (Switzerland). 2012;**12**(6):7207-7258

[29] Lupan O, Ursaki VV, Chai G, Chow L, Emelchenko GA, Tiginyanu IM, et al. Selective hydrogen gas nanosensor using individual ZnO nanowire with fast response at room temperature. Sensors and Actuators B: Chemical. 2010;**144**(1):56-66

[30] Wan Q, Lin CL, Yu XB, Wang TH. Room-temperature hydrogen storage characteristics of ZnO nanowires. Applied Physics Letters. 2004;**84**(1):124

[31] Dong K-Y, Choi J-K, Hwang I-S, Lee J-W, Kang BH, Ham D-J, et al. Enhanced H2S sensing characteristics of Pt doped SnO2 nanofibers sensors with micro heater. Sensors and Actuators B: Chemical. 2011;**157**(1):154-161

[32] Zhang N, Yu K, Li Q, Zhu ZQ, Wan Q. Room-temperature high-sensitivity H[sub 2]S gas sensor based on dendritic ZnO nanostructures with macroscale in appearance. Journal of Applied Physics. 2008;**103**(10):104305

[33] Baratto C, Sberveglieri G, Onischuk A, Caruso B, di Stasio S. Low temperature selective NO2 sensors by nanostructured fibres of ZnO. Sensors and Actuators B: Chemical. 2004;**100**(1-2):261-265

[34] Sadek AZ, Choopun S, Wlodarski W, Ippolito SJ, Kalantar-zadeh K. Characterization of ZnO nanobelt-based gas sensor for H_2, NO_2 and hydrocarbon sensing. IEEE Sensors Journal. 2007;7(6):919-924

[35] Wang HT, Kang BS, Ren F, Tien LC, Sadik PW, Norton DP, et al. Hydrogen-selective sensing at room temperature with ZnO nanorods. Applied Physics Letters. 2005;**86**(24):243503

[36] Huang MH, Wu Y, Feick H, Tran N, Weber E, Yang P. Catalytic growth of zinc oxide nanowires by vapor transport. Advanced Materials. 2001;**13**(2):113-116

[37] Ahn M-W, Park K-S, Heo J-H, Park J-G, Kim D-W, Choi KJ, et al. Gas sensing properties of defect-controlled ZnO-nanowire gas sensor. Applied Physics Letters. 2008;**93**(26):263103

[38] Cao Y, Hu P, Pan W, Huang Y, Jia D. Methanal and xylene sensors based on ZnO nanoparticles and nanorods prepared by room-temperature solid-state chemical reaction. Sensors and Actuators B: Chemical. 2008;**134**(2):462-466

[39] Wang C, Chu X, Wu M. Detection of H2S down to ppb levels at room temperature using sensors based on ZnO nanorods. Sensors and Actuators B: Chemical. 2006;**113**(1):320-323

[40] Yang Z, Li L-M, Wan Q, Liu Q-H, Wang T-H. High-performance ethanol sensing based on an aligned assembly of ZnO nanorods. Sensors and Actuators B: Chemical. 2008;**135**(1):57-60

[41] Zhou X, Li J, Ma M, Xue Q. Effect of ethanol gas on the electrical properties of ZnO nanorods. Physica E: Low-dimensional Systems and Nanostructures. 2011;**43**(5):1056-1060

[42] Cho P-S, Kim K-W, Lee J-H. NO2 sensing characteristics of ZnO nanorods prepared by hydrothermal method. Journal of Electroceramics. 2006;**17**(2-4):975-978

[43] Lupan O, Chai G, Chow L. Novel hydrogen gas sensor based on single ZnO nanorod. Microelectronic Engineering. 2008;**85**(11):2220-2225

[44] Ra H-W, Choi K-S, Kim J-H, Hahn Y-B, Im Y-H. Fabrication of ZnO nanowires using nanoscale spacer lithography for gas sensors. Small. 2008;**4**(8):1105-1109

[45] Sultan SM, Ditshego NJ, Gunn R, Ashburn P, Chong HM. Effect of atomic layer deposition temperature on the performance of top-down ZnO nanowire transistors. Nanoscale Research Letters. 2014;**9**(1):517

[46] Donthu S, Pan Z, Myers B, Shekhawat G, Wu N, Dravid V. Facile scheme for fabricating solid-state nanostructures using e-beam lithography and solution precursors. Nano Letters. 2005;**5**(9):1710-1715

[47] Ming L, Hai-Ying Z, Chang-Xin G, Jing-Bo X, Xiao-Jun F. The research on suspended ZnO nanowire field-effect transistor. Chinese Physics B. 2009;**18**(4):1594-1597

[48] Hakim MMA, Lombardini M, Sun K, Giustiniano F, Roach PL, Davies DE, et al. Thin film polycrystalline silicon nanowire

biosensors. Nano Letters. 2012;**12**(4):1868-1872

[49] Ge H, Wu W, Li Z, Jung G-Y, Olynick D, Chen Y, et al. Cross-linked polymer replica of a nanoimprint mold at 30 nm half-pitch. Nano Letters. 2005;**5**(1):179-182

[50] Weber T, Käsebier T, Szeghalmi A, Knez M, Kley E-B, Tünnermann A. Iridium wire grid polarizer fabricated using atomic layer deposition. Nanoscale Research Letters. 2011. Available from: http://www.researchgate.net/profile/Mato_Knez/publication/51738965_Iridium_wire_grid_polarizer_fabricated_using_atomic_layer_deposition/links/0fcfd511134d1c51b7000000.pdf [Accessed: 26 May 2015]

[51] Liu X, Deng X, Sciortino P, Buonanno M, Walters F, Varghese R, et al. Large area, 38 nm half-pitch grating fabrication by using atomic spacer lithography from aluminum wire grids. Nano Letters. 2006;**6**(12):2723-2727

[52] Ditshego NMJ, Sultan SM. Top-down fabrication process of ZnO NWFETs. Journal of Nano Research. 2019;**57**:77-92

[53] Ye Z, Wong M. Characteristics of thin-film transistors fabricated on fluorinated zinc oxide. IEEE Electron Device Letters. 2012;**33**(4):549-551

[54] Yang J, Park JK, Kim S, Choi W, Lee S, Kim H. Atomic-layer-deposited ZnO thin-film transistors with various gate dielectrics. Physica Status Solidi (A) - Applications and Materials Science. 2012;**209**(10):2087-2090

[55] Chong E, Kim B, Lee S. Reduction of channel resistance in amorphous oxide thin-film transistors with buried layer. IOP Conference Series: Materials Science and Engineering. 2012;**34**:012005

[56] Su B, Chu S, Juang Y. Improved electrical and thermal stability of solution-processed Li-doped ZnO thin-film transistors. IEEE Electron Device Letters. 2012;**59**(3):700-704

[57] Choi K, Jeon S, Kim H. A comparison of Ga: ZnO and Ga: ZnO/Ag/Ga: ZnO source/drain electrodes for In–Ga–Zn–O thin film transistors. Materials Research Bulletin. 2012;**47**(10):2915-2918

[58] Chen R, Zhou W, Zhang M. Self-aligned indium–gallium–zinc oxide thin-film transistor with phosphorus-doped source/drain regions. IEEE Electron Device Letters. 2012;**33**(8):1150-1152

[59] Geng D, Kang D. High-speed and low-voltage-driven shift register with self-aligned coplanar a-IGZO TFTs. IEEE Electron Device Letters. 2012;**33**(7):1012-1014

[60] Li S, Cai Y, Han D, Wang Y. Low-temperature ZnO TFTs fabricated by reactive sputtering of metallic zinc target. Electron Devices. 2012;**59**(9):2555-2558

[61] Adamopoulos G, Thomas S, Woebkenberg PH, Bradley DDC, McLachlan MA, Anthopoulos TD. High-mobility low-voltage ZnO and Li-doped ZnO transistors based on ZrO_2 high-k dielectric grown by spray pyrolysis in ambient air. Advanced Materials. 2011;**23**(16):1894

[62] Jeong S, Lee J-Y, Lee SS, Oh S-W, Lee HH, Seo Y-H, et al. Chemically improved high performance printed indium gallium zinc oxide thin-film transistors. Journal of Materials Chemistry. 2011;**21**(43):17066

[63] Navamathavan R, Nirmala R, Lee C. Effect of NH_3 plasma treatment on the device performance of ZnO based thin film transistors. Vacuum. 2011;**85**(9):904-907

[64] Kim E, Lee K, Kim D, Parsons GN, Park K, Ihm J, et al. SiNx charge trap nonvolatile memory based on ZnO thin film transistor prepared by atomic layer deposition. AIP Conf. Proc. 2011;**151**(2011):151-152

[65] Zhang L, Li J, Zhang XW, Yu DB, Jiang XY, Zhang ZL. Glass-substrate-based high performance ZnO-TFT by using a Ta2O5 insulator modified by thin SiO2 films. Physica Status Solidi (A) - Applications and Materials Science. 2010;**207**(8):1815-1819

[66] Bong H, Lee WH, Lee DY, Kim BJ, Cho JH, Cho K. High-mobility low-temperature ZnO transistors with low-voltage operation. Applied Physics Letters. 2010;**96**(19):192115

[67] Lu A, Sun J, Jiang J, Wan Q. Low-voltage transparent electric-double-layer ZnO-based thin-film transistors for portable transparent electronics. Applied Physics Letters. 2010;**96**(4):043114

[68] Zhang L, Li J, Zhang X. Low-voltage-drive and high output current ZnO thin-film transistors with sputtering SiO2 as gate insulator. Current Applied Physics. 2010;**10**(5):1306-1308

[69] Mourey DA, Member S, Zhao DA, Jackson TN. Self-aligned-gate ZnO TFT circuits. IEEE Electron Device Letters. 2010;**31**(4):326-328

[70] Bayraktaroglu B. Microwave ZnO thin-film transistors. Electron Device Letters. 2008;**29**(9):1024-1026

[71] Zhu J, Chen H, Saraf G, Duan Z. ZnO TFT devices built on glass substrates. Journal of Electronics. 2008;**37**(9):1237-1240

[72] Wei A, Pan L, Huang W. Recent progress in the ZnO nanostructure-based sensors. Materials Science & Engineering, B: Solid-State Materials for Advanced Technology. 2011;**176**(18):1409-1421

[73] Xia Y, Yang P, Sun Y, Wu Y, Mayers B, Gates B, et al. One-dimensional nanostructures: Synthesis, characterization, and applications. ChemInform. 2003;**34**(22)

[74] Hsu C, Tsai T. Fabrication of fully transparent indium-doped ZnO nanowire field-effect transistors on ITO/glass substrates. Journal of the Electrochemical Society. 2011;**158**(2):K20-K23

[75] Peng S, Su Y, Ji L. Semitransparent field-effect transistors based on ZnO nanowire networks. IEEE Electron Device Letters. 2011;**32**(4):533-535

[76] Kang CG, Kang JW, Lee SK, Lee SY, Cho CH, Hwang HJ, et al. Characteristics of CVD graphene nanoribbon formed by a ZnO nanowire hardmask. Nanotechnology. 2011;**22**(29):295201

[77] Wang Y, Sun XW, Zhao J, Goh GKL, Chen L, Liew L-L, et al. Comparison of the hydrothermal and VPT grown zno nanowire field effect transistors. International Journal of Nanoscience. 2010;**09**(4):317-320

[78] Park WI, Kim JS, Yi GC, Lee HJ. ZnO nanorod logic circuits. Advanced Materials. 2005;**17**(11):1393

[79] Ju S, Lee K, Janes DB, Dwivedi RC, Baffour-Awuah H, Wilkins R, et al. Proton radiation hardness of single-nanowire transistors using robust organic gate nanodielectrics. Applied Physics Letters. 2006;**89**(13):073510

[80] Heo Y, Tien L, Kwon Y. Depletion-mode ZnO nanowire field-effect transistor. Applied Physics. 2004;**85**(12):2274-2276

[81] Ng HT, Han J, Yamada T, Nguyen P, Chen YP, Meyyappan M. Single crystal nanowire vertical surround-gate field-effect transistor. Nano Letters. 2004;**4**(7):1247-1252

[82] Park JY, Yun YS, Hong YS, Oh H, Kim JJ, Kim SS. Synthesis, electrical and photoresponse properties of vertically well-aligned and epitaxial ZnO nanorods on GaN-buffered sapphire substrates. Applied Physics Letters. 2005;**87**(12):1-3

[83] Wang W, Xiong HD, Edelstein MD, Gundlach D, Suehle JS, Richter CA, et al. Low frequency noise characterizations of ZnO nanowire field effect transistors. Journal of Applied Physics. 2007;**101**(4)

[84] Kim H, Park J-H, Suh M, Real Ahn J, Ju S. Horizontally aligned ZnO nanowire transistors using patterned graphene thin films. Applied Physics Letters. 2012;**100**(6):063112

[85] Das S, Kim JH, Choi HS, Park YK, Hahn YB. Interfacial and electrical properties of solution processed p-TiO_2 in heterojunction devices. Electrochemistry Communications. 2011;**13**(4):350-354

[86] Noriega R, Rivnay J, Goris L, Kälblein D, Klauk H, Kern K, et al. Probing the electrical properties of highly-doped Al:ZnO nanowire ensembles. Journal of Applied Physics. 2010;**107**(7):074312

[87] Thévenot DR, Toth K, Durst RA, Wilson GS. Electrochemical biosensors: Recommended definitions and classification. Biosensors & Bioelectronics. 2001;**16**(1-2):121-131

[88] Roy S, Gao Z. Nanostructure-based electrical biosensors. Nano Today. 2009;**4**(4):318-334

[89] Jung J, Kim SJ, Lee KW, Yoon DH, Kim Y-G, Kwak HY, et al. Approaches to label-free flexible DNA biosensors using low-temperature solution-processed InZnO thin-film transistors. Biosensors & Bioelectronics. 2014;**55**:99-105

[90] Chen K, Li B, Chen Y. Silicon nanowire field-effect transistor-based biosensors for biomedical diagnosis and cellular recording investigation. Nano Today. 2011;**6**(2):131-154

[91] Yano M, Koike K, Ogata KI, Nogami T, Tanabe S, Sasa S. Zinc oxide-based biosensors. Physica Status Solidi (C) Current Topics in Solid State Physics. 2012;**9**(7):1570-1573

[92] Liu J, Goud J, Raj PM, Iyer M, Wang ZL, Tummala RR. Real-time protein detection using ZnO nanowire/thin film bio-sensor integrated with microfluidic system. In: Proceedings—Electronic Components and Technology Conference. 2008. pp. 1317-1322

[93] Liu X, Lin P, Yan X, Kang Z, Zhao Y, Lei Y, et al. Enzyme-coated single ZnO nanowire FET biosensor for detection of uric acid. Sensors and Actuators B: Chemical. 2013;**176**:22-27

[94] Mohd Azmi MA, Tehrani Z, Lewis RP, Walker KD, Jones DR, Daniels DR, et al. Highly sensitive covalently functionalised integrated silicon nanowire biosensor devices for detection of cancer risk biomarker. Biosensors & Bioelectronics. 2014;**52**:216-224

[95] Shen Y-C, Yang C-H, Chen S-W, Wu S-H, Yang T-L, Huang J-J. IGZO thin film transistor biosensors functionalized with ZnO nanorods and antibodies. Biosensors & Bioelectronics. 2014;**54**:306-310

[96] Choi A, Kim K, Jung H-I, Lee SY. ZnO nanowire biosensors for detection of biomolecular interactions in enhancement mode. Sensors and Actuators B: Chemical. 2010;**148**(2):577-582

[97] Cui Y, Wei Q, Park H, Lieber C. Nanowire nanosensors for highly sensitive and selective detection of biological and chemical species. Science (80-). 2001;**293**(5533):1289-1292

[98] Zheng G, Patolsky F, Cui Y, Wang WU, Lieber CM. Multiplexed electrical detection of cancer markers with nanowire sensor arrays. Nature Biotechnology. 2005;**23**(10):1294-1301

[99] Lu W, Xie P, Lieber CM. Nanowire transistor performance limits and applications. IEEE Transactions on Electron Devices. 2008;**55**(11):2859-2876

[100] Li Z, Chen Y, Li X, Kamins TI, Nauka K, Williams RS. Sequence-specific label-free DNA sensors based on silicon nanowires. Nano Letters. 2004;**4**(2):245-247

[101] Bunimovich YL, Shin YS, Yeo W-S, Amori M, Kwong G, Heath JR. Quantitative real-time measurements of DNA hybridization with alkylated nonoxidized silicon nanowires in electrolyte solution. Journal of the American Chemical Society. 2006;**128**(50):16323-16331

[102] Dorvel BR, Reddy B, Go J, Duarte Guevara C, Salm E, Alam MA, et al. Silicon nanowires with high-k hafnium oxide dielectrics for sensitive detection of small nucleic acid oligomers. ACS Nano. 2012;**6**(7):6150-6164

[103] Park I, Li Z, Li X, Pisano AP, Williams RS. Towards the silicon nanowire-based sensor for intracellular biochemical detection. Biosensors & Bioelectronics. 2007;**22**(9-10):2065-2070

[104] Tarasov A, Wipf M, Bedner K, Kurz J. True reference nanosensor realized with silicon nanowires, PubMed, Langmuir. May 2012;**28**(25):9899-9905. DOI: 10.1021/la301555r

[105] Zhang G-J, Zhang L, Huang MJ, Luo ZHH, Tay GKI, Lim E-JA, et al. Silicon nanowire biosensor for highly sensitive and rapid detection of dengue virus. Sensors and Actuators B: Chemical. 2010;**146**(1):138-144

[106] Stern E, Klemic J, Routenberg D. Label-free immunodetection with CMOS-compatible semiconducting nanowires. Nature. 2007;**445**(7127):519-522

[107] Lee M-H, Lee K-N, Jung S-W, Kim W-H, Shin K-S, Seong W-K. Quantitative measurements of C-reactive protein using silicon nanowire arrays. International Journal of Nanomedicine. 2008;**3**(1):117-124

[108] Chen S, Bomer JG, van der Wiel WG, Carlen ET, van den Berg A. Top-down fabrication of sub-30 nm monocrystalline silicon nanowires using conventional microfabrication. ACS Nano. 2009;**3**(11):3485-3492

[109] Sun K, Zeimpekis I, Lombardini M, Ditshego NMJ, Pearce SJ, Kiang KS, et al. Three-mask polysilicon thin-film transistor biosensor. IEEE Transactions on Electron Devices. 2014;**61**(6):2170-2176

[110] Zhang Y, Kang Z, Yan X, Liao Q. ZnO nanostructures in enzyme biosensors. Science China Materials. 2015;**58**(1):60-76

[111] Wei A, Sun XW, Wang JX, Lei Y, Cai XP, Li CM, et al. Enzymatic glucose biosensor based on ZnO nanorod array grown by hydrothermal decomposition. Applied Physics Letters. 2006;**89**(12):123902

[112] Yang K, She G-W, Wang H, Ou X-M, Zhang X-H, Lee C-S, et al. ZnO nanotube arrays as biosensors for glucose. Journal of Physical Chemistry C. 2009;**113**(47):20169-20172

[113] Aydoğdu G, Zeybek DK, Pekyardımcı Ş, Kılıç E. A novel amperometric biosensor based on ZnO

nanoparticles-modified carbon paste electrode for determination of glucose in human serum. Artificial Cells, Nanomedicine, and Biotechnology. 2013;**41**(5):332-338

[114] Fulati A, Ali SMU, Asif MH, Alvi NH, Willander M, Brännmark C, et al. An intracellular glucose biosensor based on nanoflake ZnO. Sensors and Actuators B: Chemical. 2010;**150**(2):673-680

[115] Yang Z, Zong X, Ye Z, Zhao B, Wang Q, Wang P. The application of complex multiple forklike ZnO nanostructures to rapid and ultrahigh sensitive hydrogen peroxide biosensors. Biomaterials. 2010;**31**(29):7534-7541

[116] Wang JX, Sun XW, Wei A, Lei Y, Cai XP, Li CM, et al. Zinc oxide nanocomb biosensor for glucose detection. Applied Physics Letters. 2006;**88**(23):233106

[117] Liu J, Guo C, Li CM, Li Y, Chi Q, Huang X, et al. Carbon-decorated ZnO nanowire array: A novel platform for direct electrochemistry of enzymes and biosensing applications. Electrochemistry Communications. 2009;**11**(1):202-205

[118] Yang C, Xu C, Wang X. ZnO/Cu nanocomposite: A platform for direct electrochemistry of enzymes and biosensing applications. Langmuir. 2012;**28**(9):4580-4585

[119] Wei Y, Li Y, Liu X, Xian Y, Shi G, Jin L. ZnO nanorods/Au hybrid nanocomposites for glucose biosensor. Biosensors & Bioelectronics. 2010;**26**(1):275-278

[120] Chu X, Zhu X, Dong Y, Chen T, Ye M, Sun W. An amperometric glucose biosensor based on the immobilization of glucose oxidase on the platinum electrode modified with NiO doped ZnO nanorods. Journal of Electroanalytical Chemistry. 2012;**676**:20-26

[121] Curreli M, Zhang R, Ishikawa FN, Chang HK, Cote RJ, Zhou C, et al. Real-time, label-free detection of biological entities using nanowire-based FETs. IEEE Transactions on Nanotechnology. 2008;**7**(6):651-667

[122] Wang WU, Chen C, Lin KH, Fang Y, Lieber CM. Labelfree detection of small-molecule–protein interactions by using nanowire nanosensors. Proceedings of the National Academy of Sciences of the United States of America. 2005;**102**:3208-3212

[123] Lei Y, Yan X, Luo N, Song Y, Zhang Y. ZnO nanotetrapod network as the adsorption layer for the improvement of glucose detection via multiterminal electron-exchange. Colloids and Surfaces A: Physicochemical and Engineering Aspects. 2010;**361**(1-3, 173):169

[124] Zhang F, Wang X, Ai S, Sun Z, Wan Q, Zhu Z, et al. Immobilization of uricase on ZnO nanorods for a reagentless uric acid biosensor. Analytica Chimica Acta. 2004;**519**(2):155-160

[125] Ali SMU, Ibupoto ZH, Kashif M, Hashim U, Willander M. A potentiometric indirect uric acid sensor based on ZnO nanoflakes and immobilized uricase. Sensors (Basel). 2012;**12**(3):2787-2797

[126] Jindal K, Tomar M, Gupta V. Inducing electrocatalytic functionality in ZnO thin film by N doping to realize a third generation uric acid biosensor. Biosensors & Bioelectronics. 2014;**55**:57-65

[127] Giri AK, Sinhamahapatra A, Prakash S, Chaudhari J, Shahi VK, Panda AB. Porous ZnO microtubes with excellent cholesterol sensing and catalytic properties. Journal of Materials Chemistry A. 2013;**1**(3):814-822

[128] Ahmad R, Tripathy N, Hahn Y-B. Wide linear-range detecting high

sensitivity cholesterol biosensors based on aspect-ratio controlled ZnO nanorods grown on silver electrodes. Sensors and Actuators B: Chemical. 2012;**169**:382-386

[129] Ahmad M, Pan C, Luo Z, Zhu J. A single ZnO nanofiber-based highly sensitive amperometric glucose biosensor. Journal of Physical Chemistry C. 2010;**114**(20):9308-9313

[130] Zhao ZW, Chen XJ, Tay BK, Chen JS, Han ZJ, Khor KA. A novel amperometric biosensor based on ZnO:Co nanoclusters for biosensing glucose. Biosensors & Bioelectronics. 2007;**23**(1):135-139

[131] Shukla SK, Deshpande SR, Shukla SK, Tiwari A. Fabrication of a tunable glucose biosensor based on zinc oxide/chitosan-graft-poly(vinyl alcohol) core-shell nanocomposite. Talanta. 2012;**99**:283-287

[132] Karuppiah C, Palanisamy S, Chen S-M, Veeramani V, Periakaruppan P. Direct electrochemistry of glucose oxidase and sensing glucose using a screen-printed carbon electrode modified with graphite nanosheets and zinc oxide nanoparticles. Microchimica Acta. 2014;**181**(15-16):1843-1850

[133] Wang Y-T, Yu L, Zhu Z-Q, Zhang J, Zhu J-Z, Fan C. Improved enzyme immobilization for enhanced bioelectrocatalytic activity of glucose sensor. Sensors and Actuators B: Chemical. 2009;**136**(2):332-337

[134] Palanisamy S, Cheemalapati S, Chen SM. Enzymatic glucose biosensor based on multiwalled carbon nanotubes-zinc oxide composite. International Journal of Electrochemical Science. 2012;**7**(9):8394-8407

Chapter 2

Metal Oxide Nanowires as Building Blocks for Optoelectronic Devices

Andreea Costas, Nicoleta Preda, Camelia Florica and Ionut Enculescu

Abstract

Metal oxide nanowires have become the new building blocks for the next generation optoelectronic devices due to their specific features such as quantum confinement and high aspect ratio. Thus, they can be integrated as active components in diodes, field effect transistors, photodetectors, sensors, solar cells and so on. ZnO, a n-type semiconductor with a direct wide band gap (3.3 eV) and CuO, a p-type semiconductor with a narrow band gap (1.2–1.5 eV), are two metal oxides which were recently in the spotlight of the researchers for applications in the optoelectronic devices area. Therefore, in this chapter we focused on ZnO and CuO nanowires, the metal oxides nanowire arrays being prepared by straightforward wet and dry methods. Further, in order to emphasize their intrinsic transport properties, lithographic and thin films deposition techniques were used to integrate single ZnO and CuO nanowires into diodes and field effect transistors.

Keywords: metal oxide nanowire arrays, single ZnO and CuO nanowires, lithographic techniques, diodes, field effect transistors

1. Introduction

Over the last decades, metal oxide nanowires, one dimensional nanostructures characterized by a high aspect ratio [1], have gained a special interest owed among others to their large specific area given by the size effects. This feature is responsible for their high sensitivity that is very important in a wide range of applications in optoelectronics [2], electrochemical sensors [3], spintronics [4], photocatalysis [5], noninvasive medical diagnosis [6], drug delivery [7], etc. Thus, due to their high sensitivity, metal oxide nanowires can detect even a single molecule, or even mechanical, optical or electrical signals [8–10]. For example, the size of biological molecules, such as proteins and nucleic acids, is comparable to the size of nanostructures, therefore any interaction between these molecules should induce major changes in the properties of the nanowires. Consequently, the metal oxide nanowires can be regarded as the perfect candidates for integration as single components in diodes [11], field effect transistors [12, 13], advanced biosensors [14], photodetectors [15], light emitting diodes [16], solar cells [17], magnetoresistive sensors [18], etc.

The preparation methods represent the key factor in order to obtain metal oxide nanowires with tunable dimensions and tailored physico-chemical properties. To date, many preparation approaches were used for preparing arrays of metal oxide nanowires, such as template electrodeposition [19], electroless deposition [20], sol–gel [21], chemical bath deposition [22], hydrothermal growth [23], pulsed laser deposition [24], chemical vapor deposition [25], atomic layer deposition [26], lithographic techniques [27], etc. Among them, chemical synthesis carried out in water as a reaction medium is a simple wet preparation route, easy to process and suitable for large-scale synthesis of metal oxide powders consisting in micro- and nano-structures with different morphologies featured by a good crystallinity [13, 28, 29]. Another preparation method, thermal oxidation in air is a relatively facile dry, low-cost, non-hazardous and high throughput approach that can be used at a large-scale to obtain metal oxides nanostructures of a high purity and crystallinity [13]. Using thermal oxidation in air to obtain arrays of metal oxide nanowires, the nanowires length, diameter and density can be easily controlled by modifying the parameters involved in the thermal oxidation in air process, such as: the heating rate, the annealing temperature and the time of the treatment [30].

Zinc oxide (ZnO) is an interesting eco-friendly and versatile metal oxide, suitable for many applications due to its remarkable physico-chemical properties. Being an n-type semiconductor with a direct band gap (3.37 eV), with an excitonic binding energy of 60 meV, it can be easily integrated in optoelectronic devices such as photodetectors [15], light emitting diodes [31], solar cells [32]. Additionally, its flexibility in terms of nanostructure morphology (nanowires, nanotubes, nanofibers, nanorods, nanoneedles, hexagonal nanoprisms, nanoflowers, rings, etc.) [33, 34] offers another important advantage for applications in sensors [35], photocatalysis [36], as well as in surfaces with self-cleaning properties [37].

Copper oxide (CuO) is a p-type semiconductor easy to prepare, with a high stability, having an indirect narrow band gap (1.2–1.8 eV). This metal oxide can be implemented in various applications such as: solar cells [38], field effect transistor [11], gas sensors [39], photocatalysis [40], water purification [41], etc. Being also an antiferromagnetic material below 220 K with a local magnetic moment of about 0.6 μ_B, CuO was also investigated for application in magnetic storage units [42].

In this chapter, we present our research regarding the preparation and complex characterization of metal oxide nanowire arrays by wet (chemical synthesis in aqueous solution) and dry (thermal oxidation in air) approaches. In addition, electronic devices based on single metal oxide nanowires were developed and analyzed in terms of electrical characterization. Further, lithographic techniques such as photolithography, electron beam lithography and focused ion beam induced deposition, combined with radio-frequency magnetron sputtering and thermal vacuum evaporation were used for fabricating electronic devices like diodes and field effect transistors based on single metal oxide (ZnO or CuO) nanowires. In addition, the electrical properties of the electronic devices based on single metal oxide nanowires prepared by wet and dry methods were analyzed and discussed.

2. Lithographic techniques

Lithographic techniques are mainly used in modern semiconductor manufacturing industry, in micro- and nano-fabrication to pattern thin films with specific geometries integrated into electronic devices. Lithographic techniques are subsequently divided according to the targeted application into: photolithography, electron beam lithography, focused ion beam induced deposition, extreme UV lithography, nanoimprint lithography, colloidal lithography, soft lithography, and many others. In the following, the techniques involved in the fabrication of our electronic devices based on single metal oxide nanowires are briefly described.

2.1 Photolithography

Photolithography is a conventional lithographic process in which specific geometric shapes drawn on a photomask are transferred to the desired substrate by means of light in the UV range, using a lamp that emits UV light of a certain wavelength and a polymer that has a photoactive component sensitive to the UV light named photoresist. In this case, the resolution is limited by the wavelength used and by the type of the aligner. In our case, the photolithography was involved in the fabrication of Ti/Au metallic interdigitated electrodes onto Si/SiO_2 wafers. The main steps of the process being: cleaning of the Si/SiO_2 wafers, deposition of the primer and the photoresist by centrifugation, baking of the deposited layer, the mask alignment, the UV exposure, baking after UV irradiation, the UV exposure through clear mask, the development process and the deposition of Ti and Au thin films using radio-frequency (RF) magnetron sputtering and thermal vacuum evaporation, respectively.

2.2 Electron beam lithography

Electron beam lithography (EBL) is a lithographic technique that uses a high-voltage accelerated electron beam to directly pattern a substrate without the need of a mask. In this process, the electron beam irradiates a thin layer of electron-sensitive polymer which was previously deposited on the substrate. During the exposure, the polymer (electron beam resist) bonds break and thus after the lift-off process, the appropriate geometric configuration is obtained. This technique has a high resolution, that can be used in nanoscale electronics, optoelectronics and photonics. In our case, in order to contact single nanowires using EBL, the first step is to transfer the nanowires between the metallic interdigitated electrodes. Further, a layer of electron beam resist –polymethyl methacrylate (PMMA) is deposited on the sample. Then, the desired contacts are designed into a CAD program and exposed with the electron beam. After the development process, the Ti and Au or Pt thin films are deposited by RF magnetron sputtering and thermal vacuum evaporation, respectively. Finally, the lift-off process removes the excess metal and the Ti/Au contacts are obtained.

2.3 Focused ion beam induced deposition

Focused ion beam induced deposition (FIBID) is a lithographic technique that uses a highly focused ion beam of gallium ions (Ga^+), a gas injection system and an organometallic precursor gas to deposit a metallic thin film without the need of a mask onto a substrate. In our case, similar to the EBL lithographic process, to contact single metal oxide nanowires by FIBID, the first step is to place the nanowires between the metallic interdigitated electrodes. Afterwards, the future Pt contacts are designed into a CAD program. During the deposition of the Pt contacts, an injector needle is placed very near to the substrate and upon the interaction between the organometallic compound with the ion beam, the precursor molecules are decomposed into a platinum layer and a volatile organic compound exhausted into the vacuum system. FIBID deposition is limited by the organometallic precursor gas and by the delivery rate of the gas. The deposition of a Pt contact by FIBID leads to the deposition of a carbon amorphous matrix that incorporates Pt nanoparticles implanted with Ga^+ ions.

3. Metal oxide nanowires: Influence of the preparation method type (wet or dry) on their structural, morphological and optical properties

Arrays of metal oxides (ZnO and CuO) nanowires were prepared involving two simple cost-effective wet and dry approaches: chemical synthesis in aqueous solution and thermal oxidation in air.

3.1 CuO nanowires

CuO nanowire arrays obtained by a wet method were chemically synthesized in aqueous solution based on the procedure described in Ref.s [43]. Thus, 0.0045 mol NH_4OH in 30 ml aqueous solution and 0.007 mol NaOH in 6 ml aqueous solution were added, under vigorous stirring, in a glass beaker with 100 ml aqueous solution containing 0.004 mol $CuSO_4$. The beaker was covered and stored for 7 days without stirring at ambient temperature. The precipitate was collected through centrifugation, washed several times with water and dried at room temperature.

Figure 1(a) illustrates a SEM image of the CuO nanowires chemically synthesized in aqueous solution, indicating that these nanowires have a cylindrical shape, lengths up to 2 μm and diameters of about 40 nm. The XRD pattern of the prepared CuO nanowires (**Figure 1(b)**) evidences peaks corresponding to the Miller indexes of the reflecting planes for CuO in a monoclinic phase (JCPDS reference code 00–048-1548). Based on the reflectance spectrum of the obtained CuO nanowires (**Figure 1(c)**), the band gap value was estimated as being around 1.6 eV, in agreement with previously reported data for CuO nanowires [44].

Arrays of CuO nanowire were prepared also by a dry method, using thermal oxidation in air according to the procedure given in Ref.s [11, 30]. Briefly, metallic substrates consisting in 2 cm^2 copper foils were cleaned in ultrasonic bath with acetone and isopropyl alcohol and then annealed in air for 24 h at 400°C, 500°C and 600°C in a convection oven.

The SEM images in cross-sectional view of the annealed Cu foils (**Figure 2(a), (c), (e)**) revealed that there are three distinct regions with different morphologies from the bottom up: the Cu foil, a Cu_2O thin film and the CuO nanowire arrays. Moreover, the SEM images in plan-view of the CuO nanowire arrays (**Figure 2(b), (d), (f)**) prepared by thermal oxidation in air at different temperatures disclose that the increase of the annealing temperature favors a higher density and a larger diameter of the CuO nanowires. Thus, the diameters and lengths of the CuO nanowires can be tuned as a function of the applied annealing temperatures. At 400°C, there is a low density of nanowires with diameters of

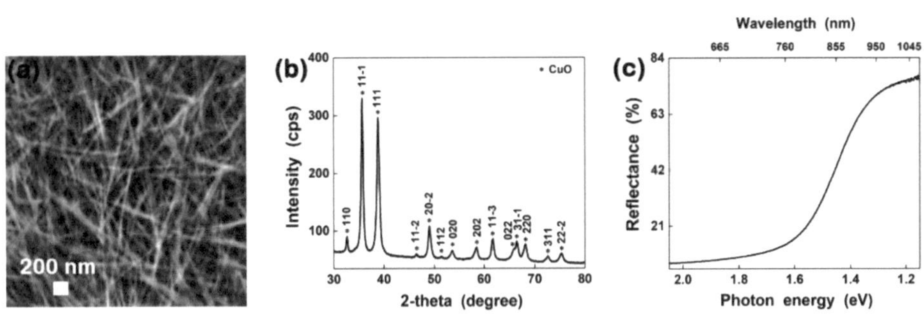

Figure 1.
(a) SEM image, (b) XRD pattern and (c) reflectance spectrum of the CuO nanowire arrays obtained by chemical synthesis in aqueous solution.

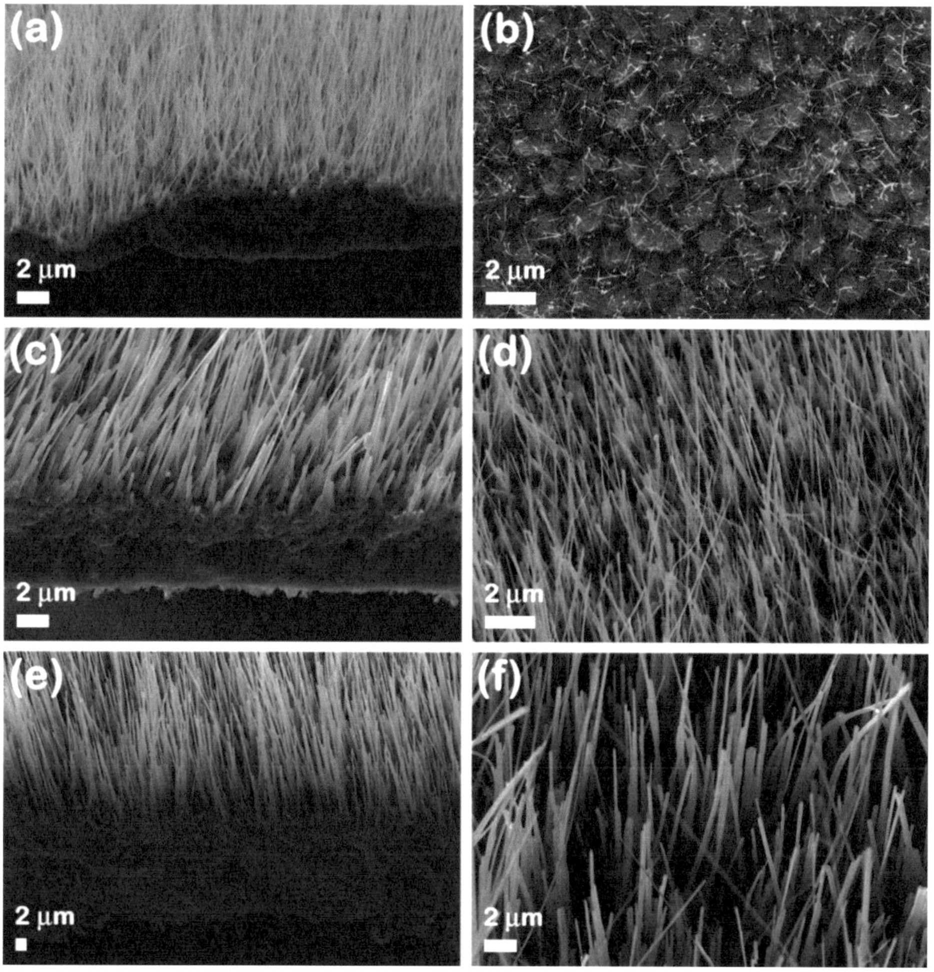

Figure 2.
SEM images of the CuO nanowire arrays prepared by thermal oxidation in air at (a), (b) 400°C, (c), (d) 500°C and (e), (f) 600°C.

about 40 nm and lengths up to 1 µm. At 500°C, there is a high density of nanowires with diameters of about 60 nm and lengths up to 30 µm. Also, at 600°C, there is a much higher density of nanowires with diameters of about 100 nm and lengths up to 30 µm.

The structural properties of the Cu foils thermally oxidized in air at different temperatures consisting in the XRD patterns (**Figure 3(a), (c), (e)**) evidence the presence of diffraction peaks assigned to the Miller indexes of the reflecting planes for three crystalline phases: Cu in face-centered-cubic phase (JCPDS reference code 00–004-0836), Cu_2O in cubic phase (JCPDS reference code 01–071-3645) and CuO in monoclinic phase (JCPDS reference code 00–048-1548). These results are in accordance with the data obtained for the CuO nanowire arrays in the cross-sectional SEM images (**Figure 2(a), (c), (e)**) in which there were clearly observed three distinct areas with different morphologies.

The band gap values for the CuO nanowire arrays obtained by thermal oxidation in air at various temperatures were assessed based on the reflectance spectra (**Figure 3(b), (d), (f)**) as being around 1.6 eV, in agreement with data previously reported in the literature for CuO nanowires [44].

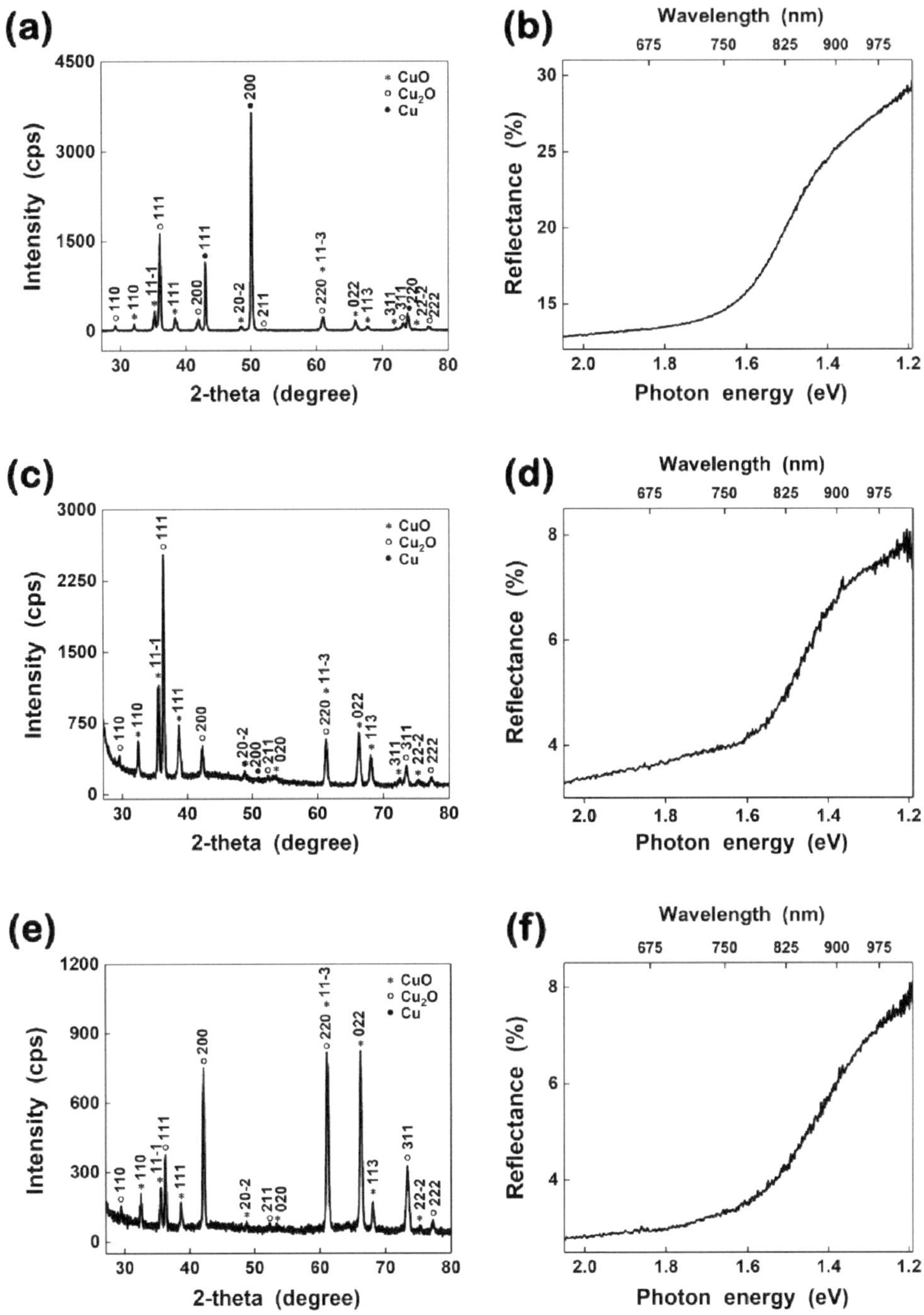

Figure 3.
(a), (c), (e) XRD patterns and (b), (d), (f) reflectance spectra of the CuO nanowire arrays prepared by thermal oxidation in air at (a), (b) 400°C, (c), (d) 500°C and (e), (f) 600°C.

3.2 ZnO nanowires

ZnO nanowire arrays were chemically synthesized in aqueous solution based on the procedures described in Ref.s [13, 28, 29]. Thus, a glass beaker with 300 ml aqueous solution containing 0.1 mmol $Zn(NO_3)_2$ and 0.1 mmol $(CH_2)_6N_4$ was covered and placed in a hot air oven, preheated at 90°C. After 5 h, the substrates, Si/SiO_2 pieces coated with a thin Ti/Au layer, were dipped and kept in the aqueous solution for 2 days. The Ti layer behaves as an adhesion promoter for the Au layer, which acts as a nucleation layer assisting the growth of ZnO nanowires.

The morphological properties of the ZnO nanowire chemically synthesized in aqueous solution are presented in **Figure 4(a)**, the SEM image revealing that the nanowires have a cylindrical shape with lengths up to 10 μm and very thin diameters of about 20 nm. The XRD pattern of the ZnO nanowires obtained by a wet method (**Figure 4(b)**) evidences peaks corresponding to the Miller indexes of the reflecting planes for ZnO crystalized in a hexagonal wurtzite phase (JCPDS reference code 00–036-1451).

The optical properties of the ZnO nanowire arrays were analyzed by reflectance and photoluminescence measurements (**Figure 4(c)** and **(d)**). From the reflectance spectrum, a band gap value was estimated of about 3.3 eV, similar with the values reported in the literature for ZnO nanowires [13]. The photoluminescence spectrum of the obtained ZnO nanowires (**Figure 4(d)**) disclose only the presence of a broad, intense emission band, centered at approximately 2.2 eV. Usually, for the ZnO nanowires synthesized in water, this broad emission band from the visible region is linked to the higher concentrations of point defects like: zinc vacancy, interstitial zinc, oxygen vacancy, interstitial oxygen, hydroxyl group, etc. [13].

Arrays of ZnO nanowire obtained by a dry route were prepared according to the method from references [13, 37]. Thus, 2 cm^2 zinc foils were cleaned in an ultrasonic bath with acetone and isopropyl alcohol and thermally oxidized in air for 24 h at 400°C, 500°C and 600°C in a convection oven.

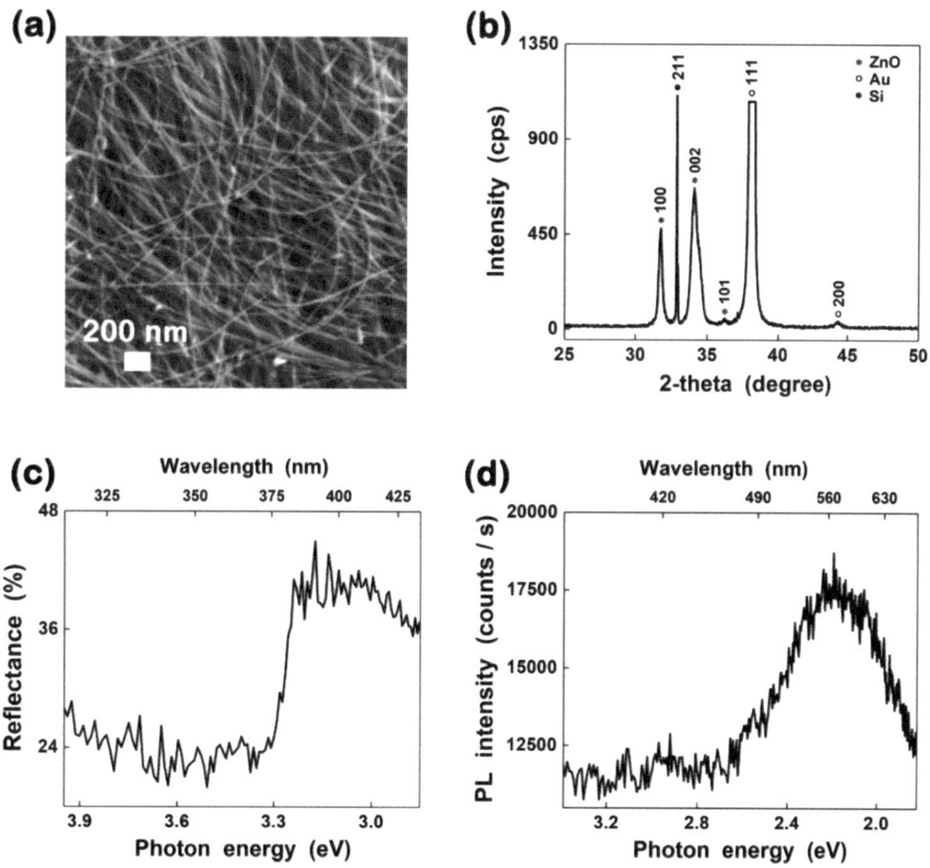

Figure 4.
(a) SEM image, (b) XRD pattern, (c) reflectance spectrum and (d) photoluminescence spectrum of the ZnO nanowire arrays obtained by chemical synthesis in aqueous solution.

The SEM images in cross-sectional view of the thermally oxidized Zn foils (**Figure 5(a), (c), (e)**) show that there are two distinct regions with different morphologies: one as a film, attributed to the metallic Zn and the second one, as nanowire arrays, associated to ZnO.

Additionally, similar with the CuO nanowire arrays obtained by a dry technique, the SEM images in plan-view of the ZnO nanowire arrays prepared by thermal oxidation in air at different temperatures (**Figure 5(b), (d), (f)**) evidence that the increase of the annealing temperature favors a higher density and a larger diameter of the ZnO nanowires. Hence, at 400°C there is a low density of nanowires with diameters of about 20 nm and lengths up to 1 μm, while at 500°C and 600°C there is a much higher density of nanowires with diameters of about 30 nm (500°C) and 60 nm (600°C) and lengths up to 30 μm.

The XRD patterns of the Zn foils thermally oxidized in air at different temperatures (**Figure 6(a), (d), (g)**) disclose the presence of diffraction peaks assigned to the Miller indexes of the reflecting planes for two crystalline phases: Zn in hexagonal phase (JCPDS reference code 00–004-0831) and ZnO crystalized in a hexagonal wurtzite phase (JCPDS reference code 00–036-1451). The structural properties are in agreement with the two distinct regions with different morphologies observed in

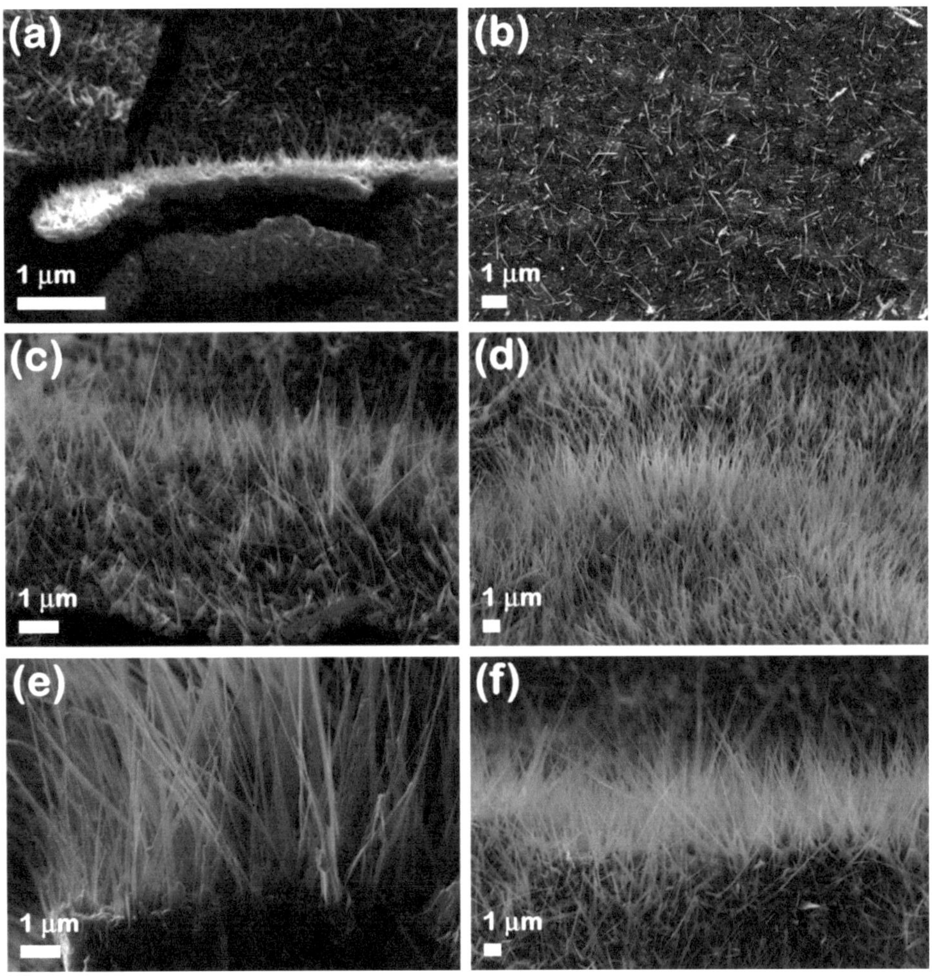

Figure 5.
SEM images of the ZnO nanowire arrays prepared by thermal oxidation in air at (a), (b) 400°C, (c), (d) 500°C and (e), (f) 600°C.

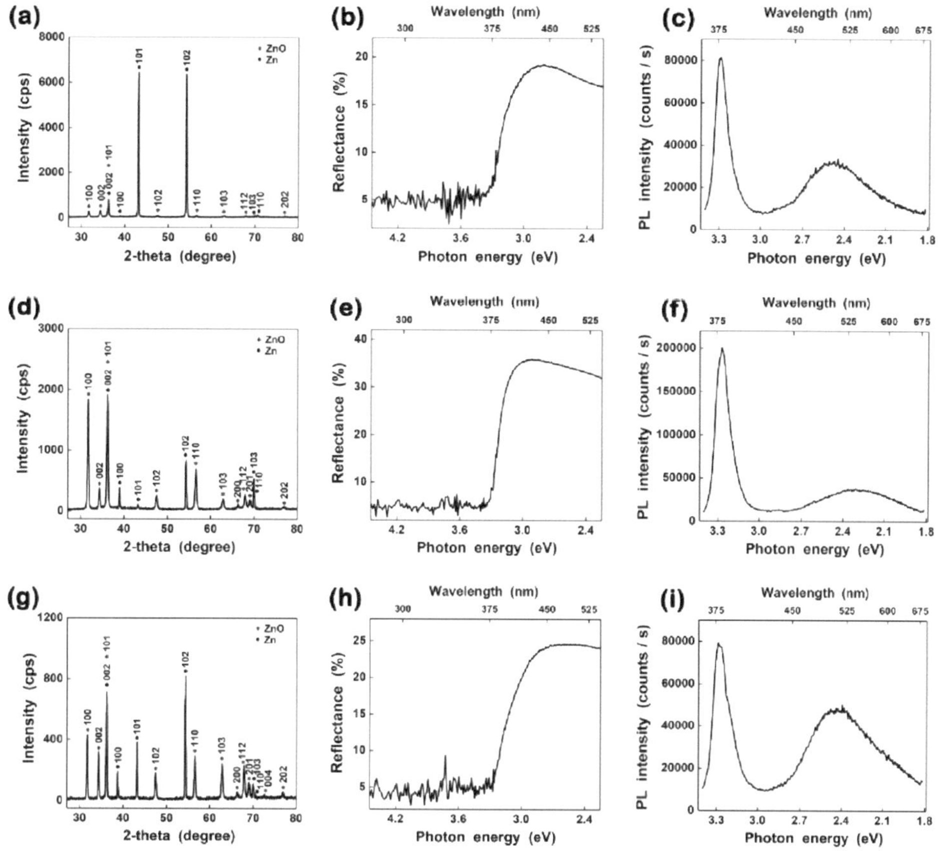

Figure 6.
(a), (d), (g) XRD patterns, (b), (e), (h) reflectance spectra and (c), (f), (i) photoluminescence spectra of the ZnO nanowire arrays prepared by thermal oxidation in air at (a), (b), (c) 400°C, (d), (e), (f) 500°C and (g), (h), (i) 600°C.

the cross-sectional SEM images (**Figure 5(a), (c), (e)**) for the ZnO nanowire arrays.

Based on the reflectance spectra of the ZnO nanowire arrays obtained by thermal oxidation in air at different temperatures (**Figure 6(b), (e), (h)**), the band gap values were estimated as being at about 3.3 eV, in accordance with data previously reports for ZnO nanowires [13].

The photoluminescence spectra of the ZnO nanowires obtained by a dry method at different annealing temperatures (**Figure 6(c), (f), (i)**) reveal the presence of two emission bands: one intense, sharp and centered at approximately 3.3 eV in the UV region and another one, weak and broad, centered at approximately 2.3 eV in the visible region. The sharp emission band in the UV region is related to the band-edge emission and the one in the visible region is linked to the various type of point defects [13].

4. Electronic devices (diodes and field effect transistors: FETs) based on single metal oxide nanowires

In order to evaluate the electrical properties and to integrate single CuO or ZnO nanowires obtained by wet (chemical synthesis in aqueous solution) and dry (thermal oxidation in air) methods into electronic devices like diodes or field effect transistors, lithographic techniques and thin films deposition techniques were employed.

Firstly, photolithography together with radio-frequency magnetron sputtering and thermal vacuum evaporation were used to pattern Si/SiO_2 wafers with Ti/Au (10 nm/100 nm) metallic interdigitated electrodes systems. Subsequently, suspensions of CuO or ZnO nanowires in ultrapure isopropyl alcohol are prepared by ultrasonication and then drop-cast onto the Si/SiO_2 substrates patterned with Ti-Au metallic interdigitated electrodes. Afterwards, single CuO or ZnO nanowires are contacted by EBL, FIBID or EBL combined with FIBID. **Figure 7(a)** illustrates a SEM image of a Si/SiO_2 substrate patterned with Ti/Au metallic interdigitated electrode system, while **Figure 7(b)** presents a SEM image of Ti/Au metallic interdigitated electrodes having metal oxide nanowires, transferred by drop-casting, between the Ti/Au electrodes.

The electrical measurements of the single CuO or ZnO nanowires contacted by lithographic techniques were carried out at room temperature in a typical two-points configuration for diodes and a three-points configuration for FETs.

4.1 Diodes and FETs based on single CuO nanowires

In the following, CuO nanowires prepared by thermal oxidation in air at 500°C were used to develop diodes and FETs based on single CuO nanowires, these nanowires being chosen owed to the smaller diameter of about 60 nm and the lengths of about 30 μm for the nanowires.

Single CuO nanowire based Schottky diodes were fabricated by contacting single CuO nanowires at one end with Ti/Au by EBL and at the other end with Pt using FIBID. This outcome can be explained considering that a Pt–CuO nanowire structure exhibits an Ohmic behavior, and that a CuO nanowire–Ti/Au structure discloses a Schottky rectifying behavior [11]. Thus, **Figure 8(a)** presents a SEM image of a single CuO nanowire prepared by thermal oxidation in air and contacted at one end with Ti/Au (100 nm/300 nm) by EBL, RF magnetron sputtering and thermal vacuum evaporation and at the other end with Pt (300 nm) by FIBID, evidencing a distance between the metallic contacts of about 11 μm. **Figure 8(b)** displays the EDX elemental mapping analysis of the metallic electrodes that contact

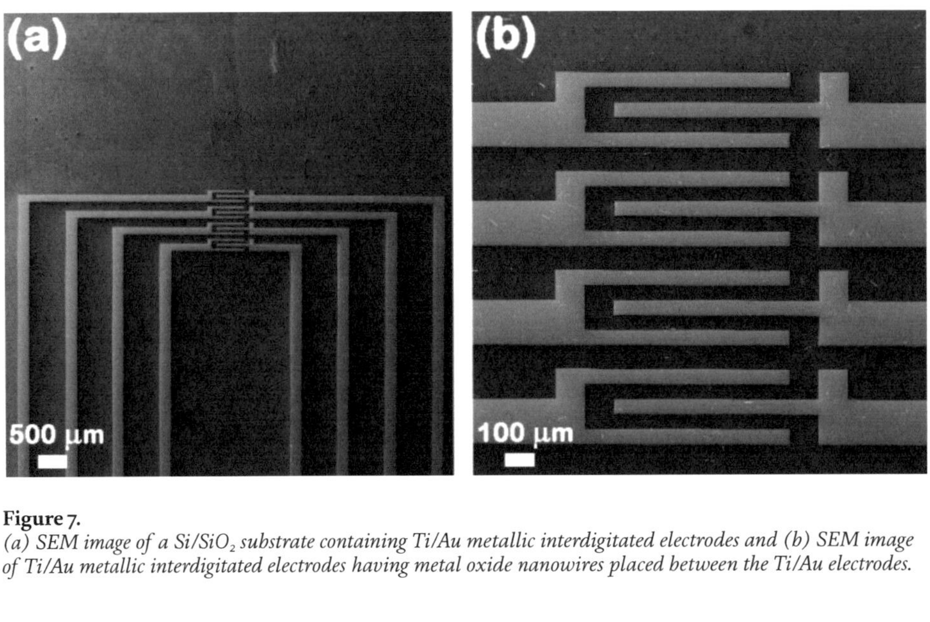

Figure 7.
(a) SEM image of a Si/SiO_2 substrate containing Ti/Au metallic interdigitated electrodes and (b) SEM image of Ti/Au metallic interdigitated electrodes having metal oxide nanowires placed between the Ti/Au electrodes.

the nanowire to the metallic interdigitated electrodes, confirming the presence of Ti and Au elements in the case of EBL and Pt for FIBID. The current–voltage characteristics of a single CuO nanowire contacted by EBL and FIBID (**Figure 8(c)**) reveal a rectifying Schottky behavior, typical for a Schottky diode [11]. Moreover, the values of the specific parameters for diodes were estimated from the current–voltage dependence to be: I_{ON}/I_{OFF} ratio $\approx 10^3$ and the ideality factor n $\approx$ 1.8, being in accordance with data reported in the literature [11].

In order to develop FETs based on single CuO nanowires, FIBID was used to contact single CuO nanowires at both ends of the nanowire with Pt (300 nm), the SEM image of a single CuO nanowire contacted by this technique being illustrated in **Figure 9(a)**. The current–voltage characteristic of a single CuO nanowire contacted by FIBID (**Figure 9(b)**) evidences a linear dependence, indicating an Ohmic contact formed between the CuO nanowire and Pt electrodes.

Figure 10(a) and **(b)** present a SEM image and the corresponding EDX mapping, proving the presence of the Pt element into the source and drain electrodes and Cu in the CuO nanowire. The length of the p-type semiconductor channel between the source and drain is about 8 μm. The output characteristics (**Figure 10(c)**) exhibit an increase in the source-drain current towards higher negative gate voltages, typical for a p-type semiconductor channel. Also, it can be noticed a change in the shape of the output characteristic at −12 V applied gate voltage, indicating the saturation region of the FET. The semilogarithmic plot of the transfer characteristic of the single CuO nanowire based FET (**Figure 10(d)**) disclose an I_{ON}/I_{OFF} ratio $\approx 10^3$, in agreement with data reported in the literature for FETs based on single nanowires [11].

4.2 Diodes and FETs based on single ZnO nanowires

In the fabrication of diodes and FETs based on single ZnO nanowires were used ZnO nanowires prepared by thermal oxidation in air at 500°C, these being chosen due to their smaller diameter of about 30 nm and the lengths of about 30 μm and also ZnO nanowires chemically synthesized in aqueous solution.

In order to develop diodes based on ZnO nanowires obtained by a dry method, single ZnO nanowire were contacted at the both ends of the nanowire Ti/Au by EBL. Hence, **Figure 11(a)** and **(b)** displays two SEM images, at different magnifications, of a single ZnO nanowire prepared by thermal oxidation in air and contacted at both ends with Ti/Pt (100 nm/200 nm) by EBL and RF magnetron sputtering, evidencing a distance between the metallic contacts of about 2 μm. The current–voltage measurement of a single ZnO nanowire contacted by EBL (**Figure 11(c)**) exhibits a non-liniar symmetrical shape indicating a structure having two back-to-back Schottky diodes, similar type of behavior being reported in the literature [45].

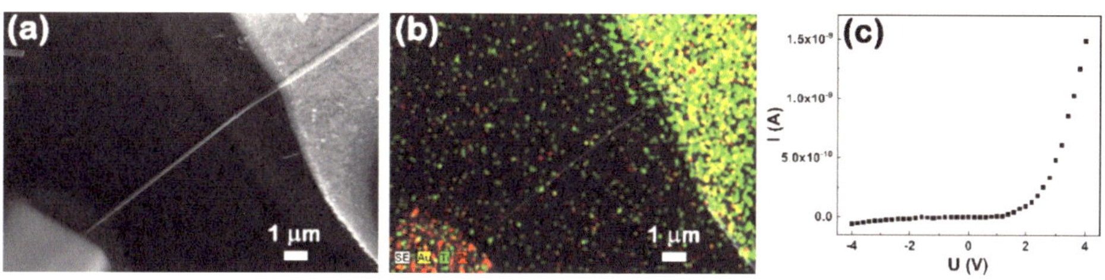

Figure 8.
(a) SEM image, (b) EDX elemental mapping and (c) current–voltage characteristic of a single CuO nanowire prepared by a dry method and contacted by EBL and FIBID.

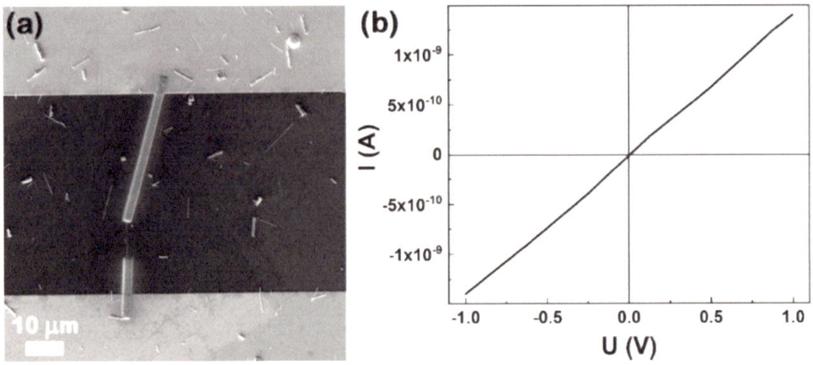

Figure 9.
(a) SEM image and (b) current–voltage of a single CuO nanowire prepared by a dry method and contacted by FIBID.

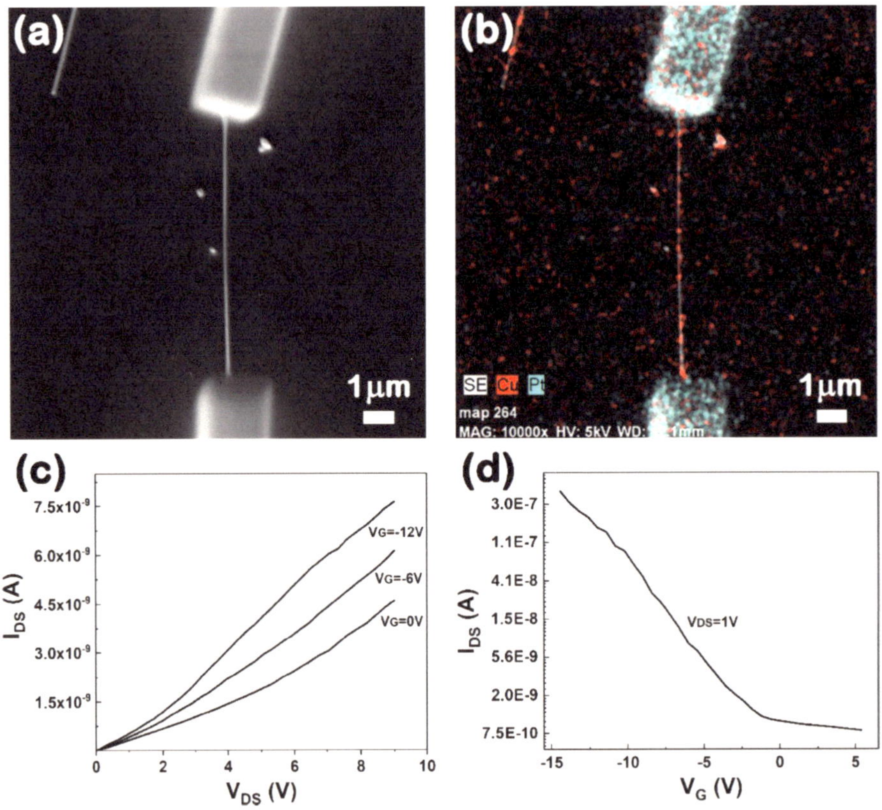

Figure 10.
(a) SEM image, (b) EDX elemental mapping, (c) output and (d) transfer characteristics of a FET based on a single CuO nanowire prepared by a dry method and contacted by FIBID.

FETs based on single ZnO nanowires prepared by a dry method were fabricated by contacting single ZnO nanowires at both ends with Ti/Au (100 nm/300 nm) by EBL, RF magnetron sputtering and thermal vacuum evaporation, the SEM image of a single ZnO nanowire contacted by EBL being illustrated in **Figure 12(a)**. The current–voltage dependence of a single ZnO nanowire contacted by EBL (**Figure 12(b)**) put in evidence a linear shape, typical for an Ohmic contact formed between the ZnO nanowire and the two Ti/Au contacts.

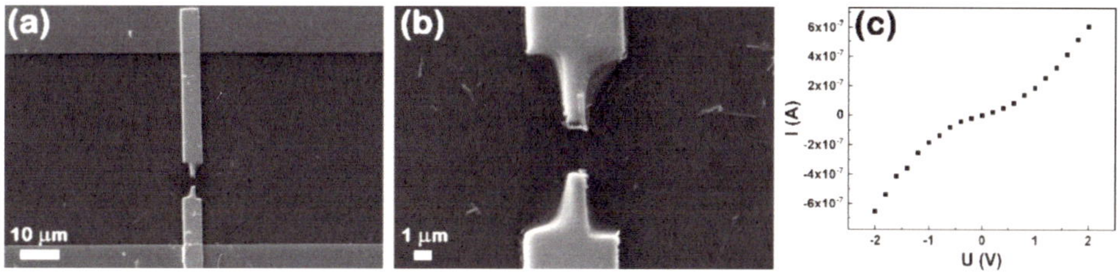

Figure 11.
(a), (b) SEM images and (c) current–voltage characteristic of a single ZnO nanowire prepared by a dry method and contacted by EBL.

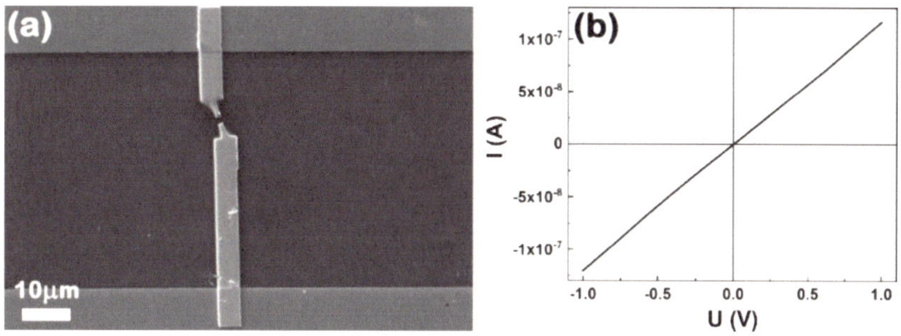

Figure 12.
(a) SEM image and (b) current–voltage characteristic of a single ZnO nanowire prepared by a dry method and contacted by EBL.

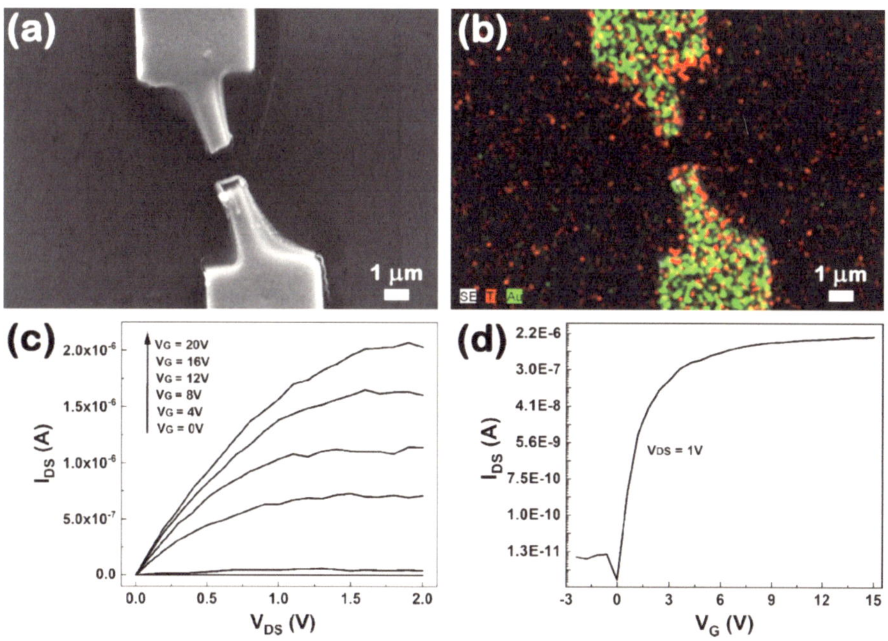

Figure 13.
(a) SEM image, (b) EDX elemental mapping, (c) output and (d) transfer characteristics of a FET based on a single ZnO nanowire prepared by a dry method and contacted by EBL.

Figure 13(a) displays a SEM image of a single ZnO nanowire contacted by EBL, indicating that the length of the n-type semiconductor channel between the source and drain is about 1 μm. The equivalent EDX mapping image (**Figure 13(b)**) proves

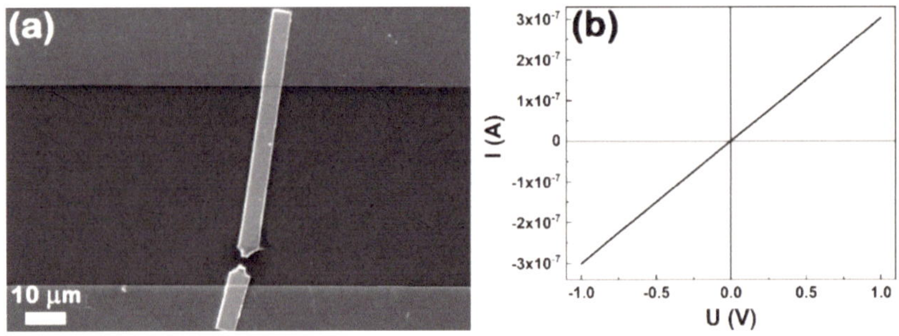

Figure 14.
(a) SEM image and (b) current–voltage characteristic of a single ZnO nanowire obtained by a wet method and contacted by EBL.

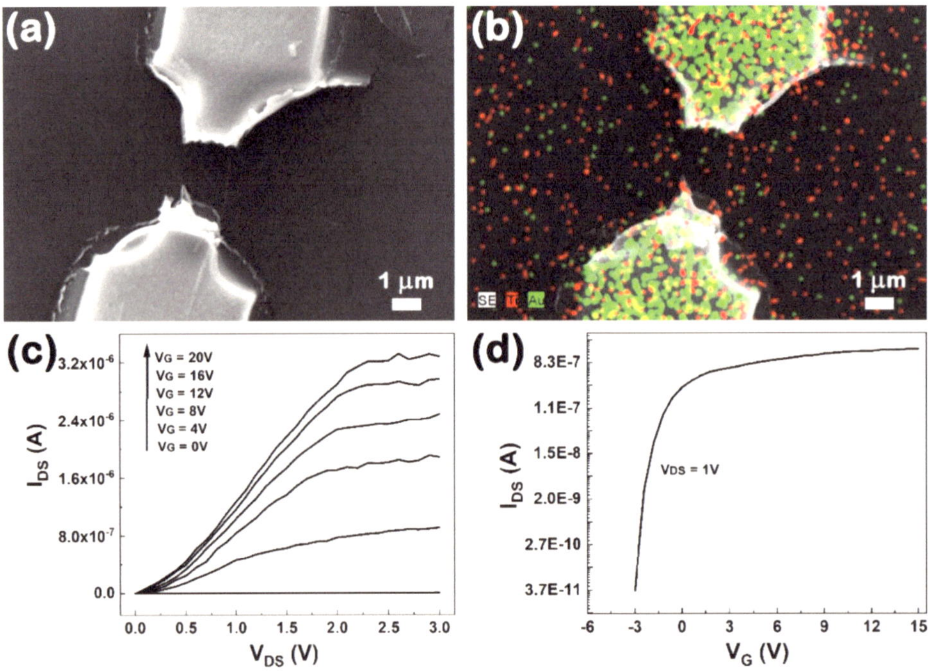

Figure 15.
(a) SEM image, (b) EDX elemental mapping, (c) output and (d) transfer characteristics of a FET based on a single ZnO nanowire obtained by a wet method and contacted by EBL.

the presence of the Ti and Au elements into the source and drain electrodes provided by EBL.

The output characteristics of a FET based on a single ZnO nanowire (**Figure 13(c)**) evidence an increase in the source-drain current towards higher positive gate voltages, typical for an n-type semiconductor channel. It can be observed that at 1.5 V applied drain-source voltage, the FET reaches the saturation region. The semilogarithmic plot of the transfer characteristic of a FET having a single ZnO nanowire as a channel (**Figure 13(d)**) exhibits an I_{ON}/I_{OFF} ratio $\approx 10^5$, in accordance with data reported in the literature for FETs based on single nanowires [13].

In the following, single ZnO nanowires synthesized by a wet technique were used as n-type channels into FET devices. Accordingly, FETs based on single ZnO nanowires were developed by contacting single ZnO nanowires at both ends with Ti/Au (100 nm/200 nm) by EBL, RF magnetron sputtering and thermal vacuum evaporation. **Figure 14(a)** reveals the SEM image of a single ZnO nanowire synthesized by a wet method contacted by EBL. Similar to the ZnO nanowires prepared by a dry path, the current–voltage characteristic of a single ZnO nanowire obtained by a wet approach contacted by EBL (**Figure 14(b)**) exhibits a linear shape, with an Ohmic contact between the ZnO nanowire and the two Ti/Au contacts.

The SEM image at a higher magnification of a single ZnO nanowire contacted by EBL (**Figure 15(a)**) evidences that the length of the nanowire channel between the source and drain is about 2 μm. The corresponding EDX mapping image (**Figure 15(b)**) confirms the presence of the Ti and Au elements into the source and drain metallic electrodes fabricated by EBL.

The output characteristics of a FET based on a single ZnO nanowire prepared by a wet method (**Figure 15(c)**) disclose an increase in the source-drain current towards higher positive gate voltages, demonstrating that the channel is an n-type semiconductor. In addition, it can be noticed that at 2 V applied drain-source voltage, the FET reaches the saturation region. The semilogarithmic plot of the transfer characteristic of a FET having a single ZnO nanowire as an n-type channel (**Figure 15(d)**) reveals an I_{ON}/I_{OFF} ratio $\approx 10^4$, the value being in accordance with data reported in the literature for ZnO single nanowires based FETs [13].

5. Conclusions

Metal oxide, ZnO and CuO, nanowire arrays were obtained by using two straightforward and cost-effective preparation methods: chemical synthesis in aqueous solution (wet) and thermal oxidation in air (dry). The influence of the preparation technique on the morphological, structural and optical properties of the metal oxide nanowire arrays were investigated. Further, ZnO and CuO nanowires prepared by these wet and dry approaches were successfully integrated as active elements into electronic devices, such as Schottky diodes and FETs by using lithographic techniques (photolithography, EBL and FIBID) and thin film deposition techniques (RF magnetron sputtering and thermal vacuum evaporation). Additionally, the characteristic parameters for the diodes and FETs were estimated from the electrical measurements: $n \approx 1.8$ and I_{ON}/I_{OFF} ratio $\approx 10^3$ for diodes and I_{ON}/I_{OFF} ratio $\approx 10^4$–10^5 for the FETs.

The main advantage of fabricating electronic devices based on single metal oxide nanowires is represented by the possibility to integrate them in optoelectronic applications.

Acknowledgements

This work has been funded by the Executive Agency for Higher Education, Research, Development and Innovation Funding (UEFISCDI), Romania, Project code: Project code: PN-III-P1-1.1-PD-2019-1102 and by the Core Program, contract no. PN19-03 (contract no. 21 N/08.02.2019) supported from the Romanian Ministry of Research and Innovation.

Conflict of interest

The authors declare no conflict of interest.

Author details

Andreea Costas*, Nicoleta Preda, Camelia Florica and Ionut Enculescu
National Institute of Materials Physics, Multifunctional Materials and Structures Laboratory, Functional Nanostructures Group, Magurele, Ilfov, Romania

*Address all correspondence to: andreea.costas@infim.ro

References

[1] Pan H., Feng Y. P. Semiconductor nanowires and nanotubes: Effects of size and surface-to-volume ratio. ACS Nano. 2008;2:2410-2414. DOI: 10.1021/nn8004872

[2] Wang Z., Nabet B. Nanowire Optoelectronics. Nanophotonics. 2015;1:491-502. DOI: 10.1515/nanoph-2015-0025

[3] Prashanthi K., Thundat T. Nanowire Sensors Using Electrical Resonance. Journal of The Electrochemical Society. 2020;167:037538. DOI: 10.1149/1945-7111/ab67a2

[4] Tian Y., Bakaula S. R., Wu T. Oxide nanowires for spintronics: materials and devices. Nanoscale. 2012;4:1529-1540.

[5] Zeng Z., Yan Y., Chen J., Zan P., Tian Q., Chen P. Boosting the Photocatalytic Ability of Cu_2O Nanowires for CO_2 Conversion by MXene Quantum Dots. Advanced Functional Materials. 2019;29:1806500. DOI: 10.1002/adfm.201806500

[6] Katwal G., Rao B. M., Varghese O. K. 1D oxide nanostructures based chemical sensors for noninvasive medical diagnosis. IEEE Sensors. 2016. DOI: 10.1109/ICSENS.2016.7808404

[7] Martínez-Banderas A. I., Aires A., Quintanilla M., Holguín-Lerma J. A., Lozano-Pedraza C., Teran F. J., Moreno J. A., Perez J. E., Ooi B. S., Ravasi T., Merzaban J. S., Cortajarena A. L., Kosel J. Iron-Based Core–Shell Nanowires for Combinatorial Drug Delivery and Photothermal and Magnetic Therapy. ACS Applied Materials & Interfaces. 2019;11:43976-43988. DOI: 10.1021/acsami.9b17512

[8] Sun Y.-L., Sun S.-M., Wang P., Dong W.-F., Zhang L., Xu B.-B., Chen Q.-D., Tong L.-M., Sun H.-B. Customization of Protein Single Nanowires for Optical Biosensing. Small. 2015;11:2869-2876. DOI: 10.1002/smll.201401737

[9] Lin C.-H., Feng M.-H., Hwang C.-H., Wu J. Y.-S., Su P.-C., Lin M.-Y., Chen C.-H., Chen B.-H., Huang B.-Y., Luc M.-P., Yang Y.-S. Surface composition and interactions of mobile charges with immobilized molecules on polycrystalline silicon nanowires. Sensors and Actuators B: Chemical. 2015;211:7-16. DOI: 10.1016/j.snb.2015.01.052

[10] Lu M.-P., Hsiao C.-Y., Lai W.-T., Yang Y.-S. Probing the sensitivity of nanowire-based biosensors using liquid-gating. Nanotechnology. 2010;21:425505. DOI: 10.1088/0957-4484/21/42/425505

[11] Florica C., Costas A., Boni A. G., Negrea R., Ion L., Preda N., Pintilie L., Enculescu I. Electrical properties of single CuO nanowires for device fabrication: Diodes and field effect transistors. Applied Physics Letters. 2015;106:223501. DOI: 10.1063/1.4921914

[12] Florica C., Matei E., Costas A., Toimil Molares M. E., Enculescu I. Field effect transistor with electrodeposited ZnO nanowire channel. Electrochimica Acta. 2014;137:290-297. DOI: 10.1016/j.electacta.2014.05.124

[13] Florica C., Costas A., Kuncser A., Preda N., Enculescu I. High performance FETs based on ZnO nanowires synthesized by low cost methods. Nanotechnology. 2016;27:475303. DOI: 10.1088/0957-4484/27/47/475303

[14] Chen M., Mu L., Wang S., Cao X., Liang S., Wang Y., She G., Yang J., Wang Y., Shi W. A Single Silicon Nanowire-Based Ratiometric Biosensor for Ca^{2+} at Various Locations in a Neuron. ACS Chemical Neuroscience.

2020;11:1283-1290. DOI: 10.1021/acschemneuro.0c00041

[15] Costas A., Florica C., Preda N., Apostol N. Kuncser A., Nitescu A., Enculescu I. Radial heterojunction based on single ZnO-Cu_xO core-shell nanowire for photodetector applications. Scientific Reports. 2019;9:5553. DOI: 10.1038/s41598-019-42060-w

[16] Zhang G., Li Z., Yuan X., Wang F., Fu L., Zhuang Z., Ren F.-F., Liu B., Zhang R., Tan H. H., Jagadish C. Single nanowire green InGaN/GaN light emitting diodes. Nanotechnology. 2016;27:435205. DOI: 10.1088/0957-4484/27/43/435205

[17] Barrigón E., Hrachowina L., Borgström M. T. Light current-voltage measurements of single, as-grown, nanowire solar cells standing vertically on a substrate. Nano Energy. 2020;78:105191. DOI: 10.1016/j.nanoen.2020.105191

[18] Enculescu I., Toimil-Molares M. E., Zet C., Daub M., Westerberg L., Neumann R., Spohr R. Current perpendicular to plane single-nanowire GMR sensor. Applied Physics A. 2007;86:43-47. DOI: 10.1007/s00339-006-3738-2

[19] Boughey F. L., Davies T., Datta A., Whiter R. A., Sahonta S.-L., Kar-Narayan S. Vertically aligned zinc oxide nanowires electrodeposited within porous polycarbonate templates for vibrational energy harvesting. Nanotechnology. 2016;27:28LT02. DOI: 10.1088/0957-4484/27/28/28LT02

[20] Kim S., Lee Y., Gu A., You C., Oh K., Lee S., Im Y. Synthesis of Vertically Conformal ZnO/CuO Core–Shell Nanowire Arrays by Electrophoresis-Assisted Electroless Deposition. The Journal of Physical Chemistry C. 2014;118:7377-7385. DOI: 10.1021/jp410293j

[21] Bae C. H., Park S. M., Ahn S.-E., Oh D.-J., Kim G. T., Ha J. S. Sol–gel synthesis of sub-50 nm ZnO nanowires on pulse laser deposited ZnO thin films. Applied Surface Science. 2006;253:1758-1761. DOI: 10.1016/j.apsusc.2006.03.006

[22] VillafuerteJ.,DonatiniF.,KioseoglouJ., Sarigiannidou E., Chaix-Pluchery O., Pernot J., Consonni V. Zinc Vacancy–Hydrogen Complexes as Major Defects in ZnO Nanowires Grown by Chemical Bath Deposition. The Journal of Physical Chemistry C. 2020;124:16652-16662. DOI: 10.1021/acs.jpcc.0c04264

[23] Fra V., Beccaria M., Milano G., Guastella S., Bianco S., Porro S., Laurenti M., Stassi S., Ricciardi C. Hydrothermally grown ZnO nanowire array as an oxygen vacancies reservoir for improved resistive switching. Nanotechnology. 2020;31:374001. DOI: 10.1088/1361-6528/ab9920

[24] Guo R. Q., Nishimura J., Matsumoto M., Nakamura D., Okada T. Catalyst-free synthesis of vertically-aligned ZnO nanowires by nanoparticle-assisted pulsed laser deposition. Applied Physics A. 2008;93:843. DOI: 10.1007/s00339-008-4791-9

[25] Liang Y. Chemical vapor deposition synthesis of Ge doped ZnO nanowires and the optical property investigation. Physics Letter A. 2019;383:2928-2932. DOI: 10.1016/j.physleta.2019.06.024

[26] Sultan S. M., Ditshego N. J., Gunn R., Ashburn P., Chong H. M. H. Effect of atomic layer deposition temperature on the performance of top-down ZnO nanowire transistors. Nanoscale Research Letters. 2014;9:517. DOI: 10.1186/1556-276X-9-517

[27] Garry S., McCarthy E., Mosnier J. P., McGlynn E. Control of ZnO nanowire arrays by nanosphere lithography (NSL) on laser-produced ZnO substrates. Applied Surface Science. 2011;257:5159-5162. DOI: 10.1016/j.apsusc.2010.11.182

[28] Pachauri V., Vlandas A., Kern K., Balasubramanian K. Site-Specific Self-Assembled Liquid-Gated ZnO Nanowire Transistors for Sensing Applications. Small. 2010;6:589-594. DOI: 10.1002/smll.200900876

[29] Vayssieres L. Growth of Arrayed Nanorods and Nanowires of ZnO from Aqueous Solutions. Advanced Materials. 2003;15:464-466. DOI: 10.1002/adma.200390108

[30] Jiang X., Herricks T., Xia Y. CuO Nanowires Can Be Synthesized by Heating Copper Substrates in Air. Nano Letters. 2002;2:1333-1338. DOI: 10.1021/nl0257519

[31] Su M., Zhang T., Su J., Wang Z., Hu Y., Gao Y., Gu H., Zhang X. Homogeneous ZnO nanowire arrays p-n junction for blue light-emitting diode applications. Optics Express. 2019;27:A1207–A1215. DOI: 10.1364/OE.27.0A1207

[32] Mahmoud F. A., Ahmed N. Elaboration of ZnO nanowires by solution based method, characterization and solar cell applications. Journal of Semiconductors. 2018;39:093002. DOI: 10.1088/1674-4926/39/9/093002

[33] Athauda T. J., Neff J. G., Sutherlin L., Butt U., Ozer R.R. Systematic Study of the Structure–Property Relationships of Branched Hierarchical TiO_2/ZnO Nanostructures. ACS Applied Materials & Interfaces. 2012;4:6917-6926. DOI: 10.1021/am302061z

[34] Wei Q., Meng G., An X., Hao Y., Zhang L. Temperature-controlled growth of ZnO nanostructures: branched nanobelts and wide nanosheets. Nanotechnology. 2005;16:2561. DOI: 10.1088/0957-4484/16/11/016

[35] Zhao S., Shen Y., Zhou P., Hao F., Xu X., Gao S., Wei D., Ao Y., Shen Y. Enhanced NO_2 sensing performance of ZnO nanowires functionalized with ultra-fine In_2O_3 nanoparticles. Sensors and Actuators B: Chemical. 2020;308:127729. DOI: 10.1016/j.snb.2020.127729

[36] Florica C., Costas A., Preda N., Beregoi M., Kuncser A., Apostol N., Popa C., Socol G., Diculescu V., Enculescu I. Core-shell nanowire arrays based on ZnO and Cu_xO for water stable photocatalysts. Scientific Reports. 2019;9:17268. DOI: 10.1038/s41598-019-53873-0

[37] Florica C., Preda N., Costas A., Zgura I., Enculescu I. ZnO nanowires grown directly on zinc foils by thermal oxidation in air: Wetting and water adhesion properties. Materials Letters. 2016;170:156-159. DOI: 10.1016/j.matlet.2016.02.035

[38] Naveena D. , Logu T. , Dhanabal R. , Sethuraman K. , Bose A. C. Comparative study of effective photoabsorber CuO thin films prepared via different precursors using chemical spray pyrolysis for solar cell application. Journal of Materials Science: Materials in Electronics. 2019;30:561-572. DOI: 10.1007/s10854-018-0322-4

[39] Li D., Tang Y., Ao D., Xiang X., Wang S., Zu X. Ultra-highly sensitive and selective H_2S gas sensor based on CuO with sub-ppb detection limit. International Journal of Hydrogen Energy. 2019;44:3985-3992. DOI: 10.1016/j.ijhydene.2018.12.083

[40] Jiang D., Xue J., Wu L., Zhou W., Zhang Y., Li X. Photocatalytic performance enhancement of CuO/Cu_2O heterostructures for photodegradation of organic dyes: Effects of CuO morphology. Applied Catalysis B: Environmental. 2017;211:199-204. DOI: 10.1016/j.apcatb.2017.04.034

[41] Xu Y., Ma J., Han Y., Zhang J., Cui F., Zhao Y., Li X., Wang W. Multifunctional

CuO Nanowire Mesh for Highly Efficient Solar Evaporation and Water Purification. ACS Sustainable Chemistry & Engineering. 2019;7:5476-5485. DOI: 10.1021/acssuschemeng.8b06679

[42] Wu D., Zhang Q., Tao M. LSDA +U study of cupric oxide: Electronic structure and native point defects. Physical Review B. 2006;73:235206. DOI: 10.1103/PhysRevB.73.235206

[43] Wang W., Wang L., Shia H., Liang Y. A room temperature chemical route for large scale synthesis of sub-15 nm ultralong CuO nanowires with strong size effect and enhanced photocatalytic activity. CrystEngComm. 2012;14: 5914-5922. DOI: 10.1039/C2CE25666E

[44] Shi W., Chopra N. Surfactant-free synthesis of novel copper oxide (CuO) nanowire–cobalt oxide (Co_3O_4) nanoparticle heterostructures and their morphological control. Journal of Nanoparticle Research. 2011;13:851-868. DOI: 10.1007/s11051-010-0086-0

[45] Meng J., Li Z. Schottky-Contacted Nanowire Sensors. Advanced Materials. 2020;32:2000130. DOI: 10.1002/adma.202000130

Chapter 3

Synthesis of Nanowire Using Glancing Angle Deposition and Their Applications

Chinnamuthu Paulsamy, Pheiroijam Pooja and Heigrujam Manas Singh

Abstract

Nanowires are highly attractive for advanced nanoelectronics and nanoscience applications, due to its novel properties such as increased surface area, large aspect ratio, and increased surface scattering of electrons and phonons. The design and fabrication of nanowires array provide a great platform to overcome the challenges/limitation of its counter partner. This chapter focuses on the synthesis of metal oxide nanowire and axial heterostructure nanowire array using the Glancing angle deposition (GLAD) technique. The structural, optical and electrical properties are studied. This GLAD technique offers control over one-dimensional (1D) nanostructure growth with self-alignment capability. It is also reviewed in an effort to cover the various application in this area of optoelectronic devices and wettability applications that had been synthesized using GLAD.

Keywords: nanowire, GLAD, heterojunction, photodetectors, wettability

1. Introduction

Low-dimensional nanostructures such as zero-dimensional (0D), one-dimensional (1D) and two-dimensional (2D) have attracted enormous attention from three-dimensional or bulk structure due to the novel physical and chemical properties caused by size and quantum effects. **Figure 1** illustrates the schematic representation of electron system in bulk structure and low-dimensional nanostructures. Quantum dot are 0D nanostructures with three quantum-confined directions. Quantum plane are 2D nanostructures with one quantum-confined direction, while two unconfined directions is available for movement of particle. Bulk structures are 0D structure with no quantum-confined directions. Nanowires (NWs) are 1D nanostructures with a large aspect ratio (length/diameter), with diameters in the 1–200 nm scale and lengths ranging from some hundreds of nanometers up to several tens of micrometer. Owing to their nanoscale dimensions, they have size confinement effects, which give them novel properties as compared to the bulk materials. Nanowires have two quantum-confined directions but one unconfined direction which is available for electrical conduction. The 1D geometry on the nanoscale of the nanowires provides an increased surface area, very high density of states, diameter dependent bandgap, and increased surface scattering of electrons and phonons [1]. These anisotropic properties are advantageous in nanoelectronics,

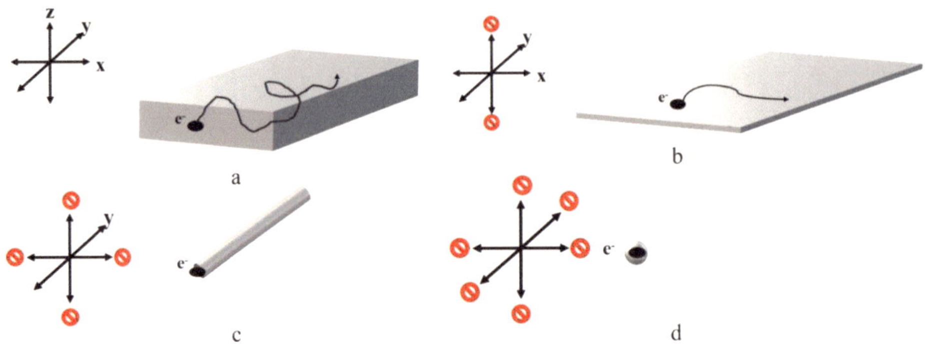

Figure 1.
Schematic illustration of electron system in (a) bulk structure, (b) quantum plane, (c) quantum wire, and (d) quantum dot.

photonics, optoelectronics, and bioengineering and have also generated great research interest [1, 2]. The concept of many advanced nanowire-based optoelectronics devices, including photodetectors, photovoltaic cells, has also been demonstrated, making nanowires promising material for advanced optoelectronics. Thus, nanowires draw considerable attention from those trying to apply nanotechnology as well as investigating in nanoscience.

Nanowires are the result of anisotropic, 1D growth on a nanometer scale. Therefore, the critical issue related to the growth of the nanowires is how to synthesize in a controlled manner. Regarding many approached have been employed, including the use of the nanolithography-based method [3], solution-based method [4], vapor-based methods [5], template-based methods [6], and glancing angle deposition (GLAD) technique [7]. Among these, the GLAD is a physical vapor deposition technique, which is a combination of oblique angle deposition and substrate positional control, which is most attractive owing to its simplicity and synthesis of different nanostructures with controlled porosity and shapes [7].

This chapter reviews the synthesis of nanowires by the GLAD technique and their applications. Furthermore, this chapter focuses on the GLAD technique. This chapter seeks to explain the understanding of the GLAD technique in the synthesis of nanowires. Accordingly, the chapter first reviews the fundamentals of the GLAD technique and the synthesis of nanowires. They are followed by some examples of optoelectronic devices and wettability applications of NWs that have been synthesized based on the GLAD technique.

2. Oblique angle deposition (OAD)

The glancing angle deposition (GLAD) technique is developed by Robbie and Brett [8], which is an extension to oblique angle deposition (OAD), where the substrate position is manipulated during the deposition. This section reviews the oblique angle deposition.

In OAD, the collimated evaporated beam is incident on the substrate surface normal with an incident angle α, as shown in **Figure 2(a)**. The incident vapor flux is treated as vector denoted by F, as shown in **Figure 2(b)** with its two components, a vertical component, and a lateral component. The arrival of incident vapor flux is a random process. During the deposition process, the impinging atoms will form islands on the substrate surface at random. The initial deposited seed will act as shadowing centers, and the tallest islands will receive more atoms as compared to the shorter ones. This phenomenon is known as the shadowing effect. With this procedure, only the tallest deposited material will grow into columns and thus result in a nanocolumn thin film formation. The lateral component is responsible for the shadowing effect. This leads to the inclination of the tilting of the nano-structure towards the direction of the incoming vapor flux, and the tilt angle is given by the tangent rule, $\tan\beta = \frac{1}{2}\tan\theta$. **Figure 3** shows how this effect takes place. In general, the columnar tilt angle β is less than the vapor flux incident angle [7].

The thin films deposited by OAD shows the following properties: Porous thin films acquiring nano-columnar structures. The nanocolumns tilt away from the substrate surface normal and towards the incident vapor flux direction.

2.1 Glancing angle deposition

The GLAD technique is developed based on the OAD with the only addition, which is the manipulation of the substrate position by using two stepper motors, one controls the incident angle, α, and the other motor controls the azimuthal rotation of the substrate with respect to the substrate surface normal. **Figure 4** shows the GLAD setup. The tilting of the nanocolumn structure found in OAD is mitigated primarily to the rotation of the substrate, which cancels out the lateral component of the incident vapor flux during the deposition process. By changing the speed and phase of the azimuthal rotation along with the deposition rate, different sculptures of nano-columns such as C-shape, S-shape, helical or vertical nanocolumns can be achieved.

The nanocolumns synthesized by GLAD show the following properties: The porosity of the film is controlled by changing the incident angle. Self-alignment is

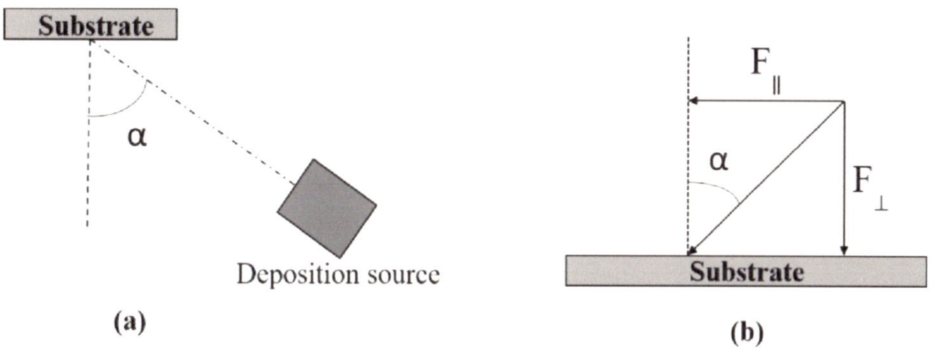

Figure 2.
(a) Experimental setup for OAD and (b) incident vapor flux components.

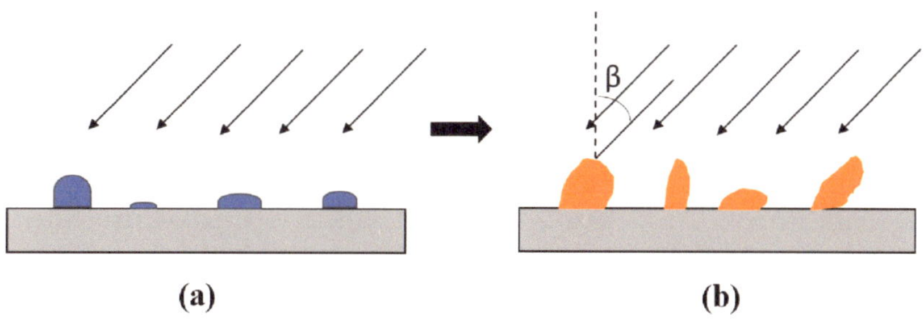

Figure 3.
The shadowing effect during OAD (a) initial nucleation to form shadowing centres, and (b) columnar structures formed due to the shadowing effect.

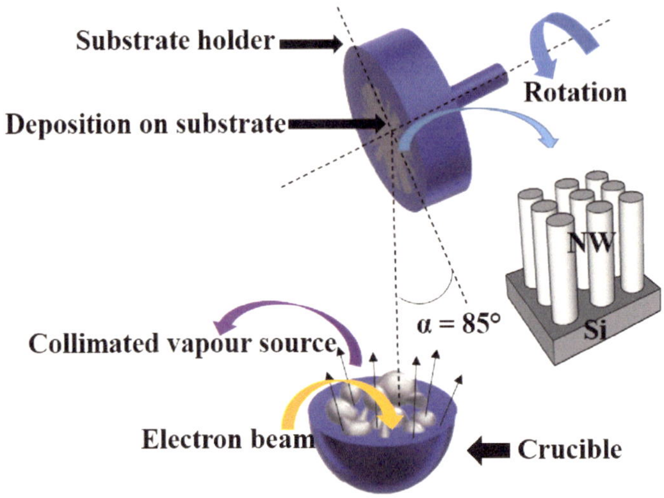

Figure 4.
Experimental setup for GLAD.

due to the shadowing effect. The shape and in-plane alignment of nanocolumns can be modified. GLAD is compatible with a large number of materials [9]. **Table 1** summarizes the synthesis of nanowires using the GLAD technique. The following sections give examples of the various applications of nanowires synthesized using the GLAD technique. The general experimental conditions are the following: the experiments were performed in an electron-beam evaporator (BC 300 HHV India) incorporated with GLAD under high vacuum pressure (6×10^{-6} mbar) inside the chamber. The deposition rate varied from 0.5–1 Å/s, and rotation speed was maintained at 20–30 rpm. The substrate holder was oriented at 85° with respect to the source. During the deposition process, the growth rate was monitored by using a quartz crystal-based digital thickness monitor inside the chamber.

2.2 Properties of nanowires synthesized using GLAD

This section discusses the structural and optical properties of nanowires. For various applications, characterizing the structural, optical, and electrical properties of nanowires is important so that the interrelationship can be investigated and established. Nanowires synthesized using the same material under the same experimental conditions may possess different properties due to the differences in structural properties.

2.2.1 Structural characterization

Structural properties of a nanowire help in determining the various attribute like optical and electrical properties. X-ray diffraction (XRD), scanning electron microscope (SEM), transmission electron microscope (TEM) are used to investigate the structural properties of nanowires. XRD characterization provides crystal structure information. The peak in XRD provides the crystal phase structure long with growth direction. SEM produces images of the nanowires down to the length scale of ~100 nm, which gives information regarding the structural arrangement, geometrical features of the nanowire. For example, SEM images of TiO_2 nanowire arrays grown on Si substrate provide evidence for shadowing effect as well as

Material	Deposition method	Reference
TiO_2	Electron beam	[10–15]
TiO_2/In_2O_3	Electron beam	[16]
SnO_2	Electron beam	[17]
WO_2	Electron beam	[18]
Er_2O_3-doped SnO_2	Electron beam	[19]
Er_2O_3	Electron beam	[20]
Er-doped TiO_2	Electron beam	[21, 22]
Co_3O_4-TiO_2	Electron beam	[23]
SiO_x-$In_{2-x}O_{3-y}$	Electron beam	[24]
Ge	Electron beam	[25]

Table 1.
Summary of nanowires synthesized using the GLAD technique.

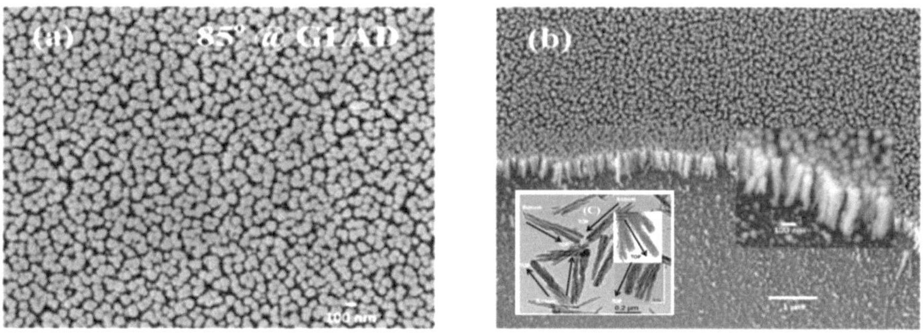

Figure 5.
FE-SEM and HR-TEM images of TiO2 nanowire: (a) top view, (b) side view, and (c) HR-TEM images (inset). Adapted from Ref. [15].

information on the geometry of the nanowires, as shown in **Figure 5(a)** and **(b)**. TEM is used for studying the nanowire at the atomic scale. Furthermore, selected area electron diffraction (SAED) patterns provide information on the crystal structure of nanowires. For example, TEM image of TiO_2 provide information on the geometry of the nanowires at nanoscale along with growth direction as shown in inset **Figure 5(c)**.

2.2.2 Optical characterization

The optical characterization of nanowires provides information on properties different from those of the bulk forms. Optical measurements like absorption and photoluminescence provide information on the absorption and emission of light in frequency range varying from UV-Vis-NIR spectra. Absorption measurement aids in determining the bandgap of the grown nanowire shown in **Figure 6**. The difference in the optical properties is due to the geometric differences such as the diameter and length of nanowires.

Photoluminescence (PL) measurement studies the optical bandgap, oxygen vacancies, and defect states in nanowires. For example, **Figure 7** investigates the PL spectrum of TiO_2 NW under 250 nm wavelength excitation.

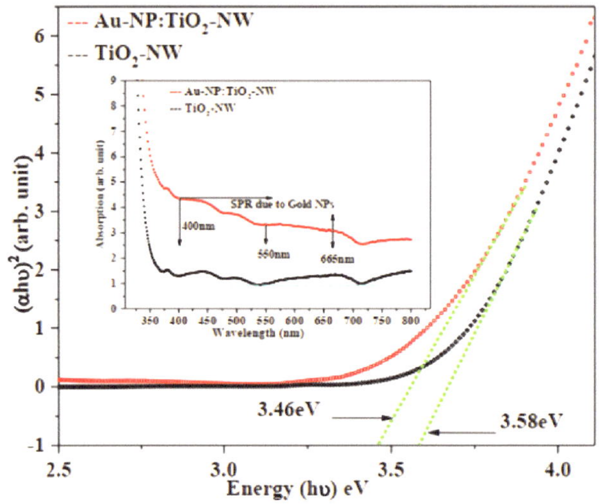

Figure 6.
Tauc plot of Au-NP:TiO_2-NW and TiO_2-NW, inset: UV-Vis absorption of both samples. Adapted from Ref. [10].

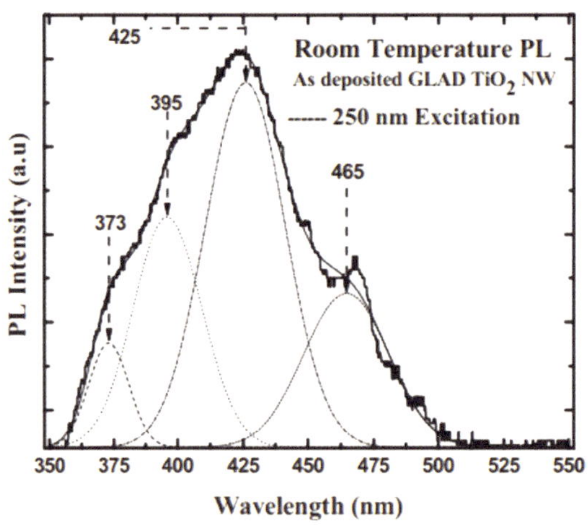

Figure 7.
Room temperature PL spectrum of as deposited TiO_2 nanowire. Adapted from Ref. [15].

3. Applications

In the preceding sections, we have discussed the central characteristics of nanowires, which attracts attention to find the application by using its novel properties compared to their bulk materials. Based on the GLAD technique, many conceptual devices have already been reported. In this section, selected applications of GLAD synthesized nanowires such as photodetectors and wettability applications are discussed. These conceptual devices were investigated based on an array of nanowires. First, the silicon (Si) substrate is subjected to a 3-step cleaning process using electronic grade acetone, methanol, and deionized (DI) water, and then an array of nanowires is synthesized on the Si substrate. The top metal contact electrode is synthesized on the nanowire arrays through an Al mask with a hole diameter ~1 mm and ITO, which is used as the back-contact electrode.

3.1 Photodetector

Semiconductor photodetectors are devices used for the detection of light. More specifically, photodetectors have applications in optical communication, flame detection, chemical, and biological detection [26]. The PN junction is one of the most commonly used configurations for semiconductor photodetectors. Many researchers have synthesized nanowire and axial heterostructure nanowire array-based photodetectors using GLAD technique. **Table 2** gives a summary of the figure of merit for the photodetectors. These photodetectors are used for the detection of UV and visible light. The performance of photodetector is evaluated by investigation of various parameters such as photosensitivity, responsivity, detectivity, and noise equivalent ratio. The photosensitivity for photodetectors is calculated from the ratio of light current to dark current given by equation below:

$$\text{Photsensitivity} = \frac{I_{\text{Photo}} - I_{\text{Dark}}}{I_{\text{Dark}}} \tag{1}$$

The responsivity at a particular wavelength is computed from the ratio of photocurrent to incident optical power defined below:

$$R_{\lambda} = \frac{I_{\text{Photo}}}{P_{\text{Opto}}} \tag{2}$$

where I_{Photo} is the photocurrent and P_{opto} is the optical power. The detectivity and NEP give the noise performance of photodetectors. Detectivity is given by

$$D^{*} = \frac{R_{\lambda}}{\sqrt{2eJ_{\text{Dark}}}} \tag{3}$$

where J_{Dark} is the dark current density, e is the charge of electron and R_{λ} is the responsivity at a particular wavelength. The NEP is expressed as

$$\text{NEP} = \frac{\sqrt{A}\sqrt{B}}{D^{*}} \tag{4}$$

where A is the area of device and B is the bandwidth. The bandwidth is assumed to be 1 kHz as the flicker noise. With these relations, detectivity and NEP are plotted as a function of voltage to evaluate the performance of photodetector. Furthermore, the photocurrent-time response gives the temporal response under light on-off to study the switching behavior at a fixed voltage. The cumulative analysis of the figure of merit for photodetectors supports that the synthesis of nanowire and axial heterostructure nanowire array-based photodetectors using the GLAD technique as a potential prospect to fulfill the requirements of commercial UV-Visible photodetectors.

Device structure	Wavelength (nm) (voltage bias)	Responsivity	Detectivity (Jones)	NEP	Rise time	Fall time	Ref.
TiO_2-NW/p-Si	370 (−7 V)	28.38 A/W	4.97×10^{11}	5.6×10^{-12} W	0.30 s	0.18 s	[10]
Au-NP:TiO_2-NW/p-Si	370 (−7 V)	43.27 A/W	2.63×10^{12}	1.0×10^{-12} W	0.15 s	0.17 s	[10]
TiO_2/In_2O_3 AH NW	380 (−4.5 V)	11.17 A/W	3.14×10^{14}	8.9×10^{-15} W	52.96 ms	89.32 ms	[16]
WO_3	360 (3 V)	9.66 A/W	5.94×10^{12}	—	1.78 s	1.09 s	[18]
		50.95 A/W (annealed)	1.05×10^{13}	—	1.33 s	0.94 s	
Ag/Er_2O_3/Si	210 (−3 V)	0.527 A/W	1:18 × 10^{12}	2.37 pW	0.28 s	0.18 s	[20]

Table 2.
Summary of the figure of merit for photodetectors.

3.2 Wettability application

In today's world, various technologies have been obtained from nature. Among them, self-cleaning technology is one of it. Many surfaces found in nature show self-cleaning properties. The leaves of plants such as lotus [27] and wings of butterflies [28] are a few examples. Self-cleaning technology obtained a lot of attention in the late 20th century for applications ranging from solar panel cleaning, windowpane cleaning, and textiles to cement. In recent years, many research works are carried out to develop durable and efficient self-cleaning coating surfaces with improved optical qualities. The self-cleaning coating can be classified into two categories: hydrophobic and hydrophilic coatings. In the hydrophobic coating technique, water droplets roll and slide over the surfaces and clean them, while in the hydrophilic technique, water forms a sheet of water over the surfaces and carries the dirt and other impurities away. The phenomenon of self-cleaning is associated with surface contact angle, which is the angle formed between the surfaces of the liquid droplet to the solid surface. Generally, when the contact angle is less than 90^0, the solid surface is defined as a hydrophilic surface (**Figure 8(a)**). If the contact angle is greater than 90^0, the surface is termed as a hydrophobic surface (**Figure 8(b)**). Likewise, the surface with a water contact angle close to zero is defined as super hydrophilic, and surface with a contact angle greater than 150^0 is categorized as super hydrophobic (**Figure 8(c)**).

Young in 1805 proposed a model to define the state of liquid droplet on an ideal rigid surface [29]. The equation defined by Young's model is given as:

$$\gamma_{SG} = \gamma_{SL} + \gamma_{LG}\cos\theta_Y \tag{5}$$

θ_Y is the water contact angle and Υ_{SG}, Υ_{SL} and Υ_{LG} are the interfacial surface tensions of solid and gas, solid and liquid, liquid and gas, respectively. This

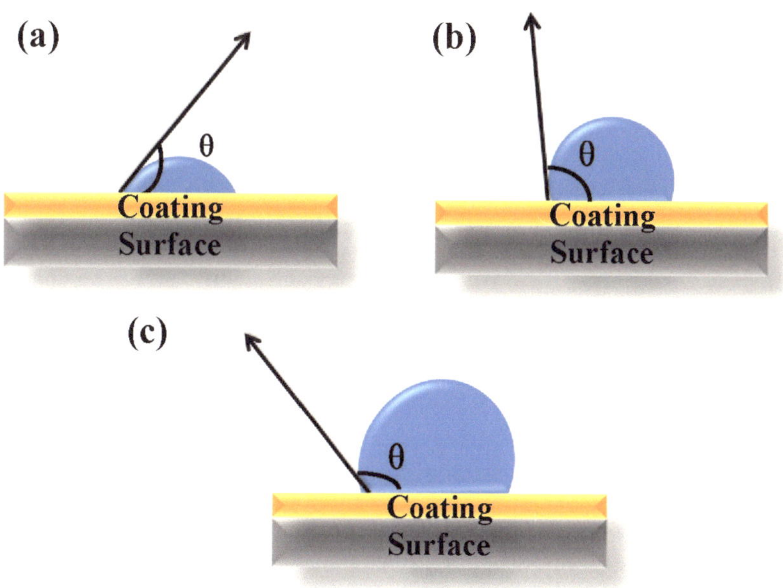

Figure 8.
Schematic diagram of (a) hydrophilic (θ < 90°), (b) hydrophobic (θ > 90°), and (c) super hydrophobic (θ > 150°) surfaces.

equation shows the water contact angle of a liquid droplet on solid surface from the three surface tensions. As seen in **Figure 9(a)**, an equilibrium state is reached between these three surface tensions, and the contact angle is given by the angle between Υ_{SL} and Υ_{LG}. In 1936, Wenzel further purposed a model to describe homogeneous wetting regimes [30]. This wetting regime for water contact angle on rough surfaces is defined by the equation given below:

$$\cos\theta_W = r\cos\theta_Y \tag{6}$$

where θ_W and θ_Y represent apparent contact angle and Young's contact angle for an ideal rigid surface respectively while r represents the surface roughness factor, which is the ratio of true solid surface area to the apparent surface area. This model states on the affiliation between the structure and surface tension of a homogeneous surface. In **Figure 9(b)**, the true surface area is larger than the apparent surface area, thus the value of r is greater than 1. Wenzel's model is applicable to surface with single chemical component and thermodynamically stable state, which limits its applications in heterogeneous surfaces. In 1944, Cassie and Baxter defined an equation to describe contact angle for composite surfaces [31].

$$\cos\theta_C = f_1\cos\theta_1 + f_2\cos\theta_2, (f_1 + f_2 = 1) \tag{7}$$

Where θ_C is apparent contact angle in Cassie–Baxter model, θ_1 and θ_2 are intrinsic contact angle of first and second components, respectively. f_1 and f_2 are apparent area fraction of first and second component, respectively. In general, if either one of surface component is air, then $\theta_2 = 180^0$.

$$\cos\theta_C = f_1\cos\theta_1 - f_2 = f_1(\cos\theta_1 + 1) - 1 \tag{8}$$

Figure 9(c) displays a water droplet on surface showing hydrophobic property due to surface composed of air and hydrophobic component.

Metal oxides such as TiO_2, SnO_2, and ZnO have been used as a self-cleaning and antifogging surface [32, 33]. Metal oxides are known to have good stability and transparency, which enable tuning wettability on application of proper radiation. Amidst them, TiO_2 is vastly studied due to its effective photocatalytic activity as it washes off dirt or decomposes organic contaminants from surfaces [34]. Moreover, TiO_2 is known for its nontoxicity, chemical, and thermal stability [16, 35, 36]. Moreover, growth of perpendicularly aligned coaxial TiO_2/In_2O_3 NW on Si substrate is problematic due to limited growth techniques. Here, coaxial TiO_2/In_2O_3 NW assembly deposited employing GLAD technique inside electron beam evaporator [37] showed comparatively faster photo-induced wettability tuning within 10 min illumination than the reported SnO_2 doped with Fe thin film deposited using spin coating [38], TiO_2 films deposited using metal-organic chemical vapor deposition process [39], grapheme films prepared using chemical vapor deposition (CVD) method [40], TiO_2 films were fabricated on stainless steel substrates via electrophoretic deposition (EPD) [41]. Based on Cassie-Baxter relation, the water contact angle of TiO_2/In_2O_3 NW was found to be 129°. **Figure 10(a)** shows coaxial TiO_2/In_2O_3 NW FESEM side view and inset of **Figure 3(b)** schematic of TiO_2/In_2O_3 NW. In 1D NW heterostructure, there is more surface area to volume ratio and allows charge carriers to flow with less scattering enabling more carriers interaction with water as compared with uniform thin films. Vertically aligned NWs also acquires

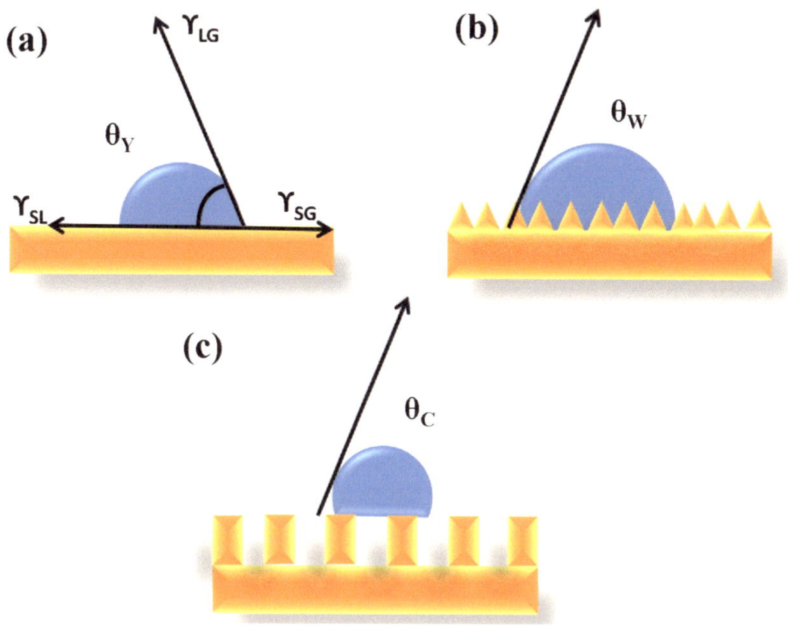

Figure 9.
(a) A liquid droplet on smooth surface in young model, (b) a liquid droplet on rough surface in Wenzel model, and (c) a liquid droplet on rough and porous surface in Cassie and Baxter model.

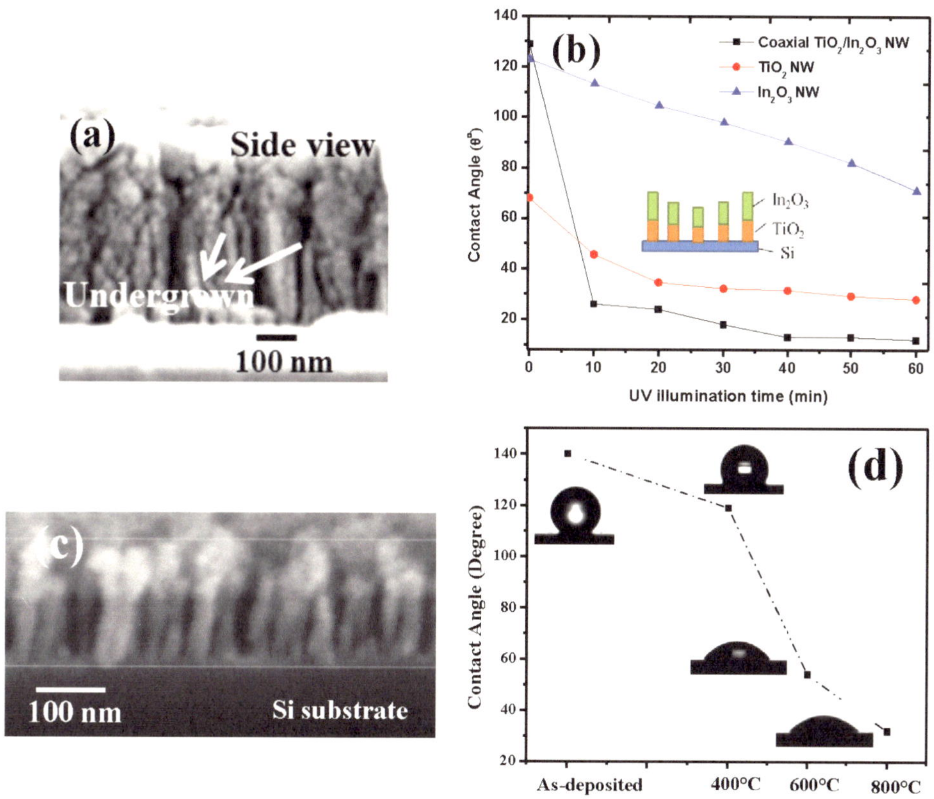

Figure 10.
(a) Coaxial TiO_2/In_2O_3 NW FESEM side view, (b) water contact angle variation under UV illumination of coaxial TiO_2/In_2O_3 NW, TiO_2 NW and In_2O_3 NW (inset schematic of TiO_2/In_2O_3 NW), (c) 600°C annealed TiO_2/In_2O_3 NW FESEM side view, and (d) static contact angle of as-deposited TiO_2-In_2O_3 NWs and annealed samples. Adapted from Ref. [37, 42].

characteristics such as low reflectivity and multiple scattering of light, which increase the carriers generation and thus more interaction of separated photogenerated charge carriers with water molecules to adsorb on the surface comparing with thin film. Moreover, TiO_2/In_2O_3 NW (1.2×10^{-3} degree^{-1} min^{-1}) showed better wettability transition rate than TiO_2 NW (3.2×10^{-4} degree^{-1} min^{-1}) and In_2O_3 NW (9.2×10^{-5} degree^{-1} min^{-1}) due to the interfacial surface modification between TiO_2 and In_2O_3 and effective interaction between photogenerated charge carriers with water molecules (**Figure 10(b)**) [37]. All these assure prospective applications of coaxial TiO_2/In_2O_3 NW grown using the GLAD technique for smart surfaces, with controlled switchable wettability by external stimuli for self-cleaning applications.

Further, TiO_2/In_2O_3 NW surface wettability had been tuned by annealing treatment, without changing the surface with extra chemical coating or by external light stimuli [42]. **Figure 10(c)** shows 600°C Annealed TiO_2/In_2O_3 NW FESEM side view. TiO_2/In_2O_3 NW samples annealed at 600°C shows nearly superhydrophilic with static water contact angle of 12° (**Figure 10(d)**). The surface of TiO_2/In_2O_3 NW had been controlled to acquire desired water contact angles, which is paramount for designing practical application in self-cleaning, electronic, and biomedical fields.

4. Conclusion

In this capture, we have reviewed the synthesis of nanowires using GLAD and their applications. We have showed that GLAD is a simple, cost effective, and catalytic free technique where well-defined vertically aligned nanostructures can be synthesized which cannot be achieved easily by nanolithography-based method, solution-based method, vapor-based methods, template-based methods. Nanowires synthesized by GLAD technique has the following advantages: Growth of vertically aligned nanostructure, the shadowing effect introduces self-alignment effect, the porosity of the nanostructure film can be controlled by changing the incident angle. Various applications of nanowire and axial heterostructure nanowire array-based photodetectors as well as wettability applications synthesized using the GLAD technique have been discussed. The performance of these applications can be improved further with different structural and growth parameters. Therefore, it can be concluded that these nanoscale-based applications have potential for future industrial and commercial applications.

Acknowledgements

The authors are grateful and would like to acknowledge, NIT Nagaland and TEQIP-III for providing the synthesis and characterization facility.

Conflict of interest

The authors declare no conflict of interest.

Author details

Chinnamuthu Paulsamy*, Pheiroijam Pooja and Heigrujam Manas Singh
Department of Electronics and Communication Engineering, NIT Nagaland, India

*Address all correspondence to: chinnamuthu@nitnagaland.ac.in

References

[1] Dresselhaus MS, Lin Y-M, Rabin O, Black MR, Kong J, Dresselhaus G. Nanowires. In: Bhushan B., editor. Springer Handbook of Nanotechnology,. Springer Handbooks. Springer; 2010. p. 119-167. DOI: 10.1007/978-3-642-02525-9_4

[2] Z. Pei L, Y. Cai Z. A Review on Germanium Nanowires. Recent Pat Nanotechnol 2011;6:44-59. DOI: 10.2174/187221012798109291

[3] Chen Y, Pepin A. Nanofabrication: Conventional and nonconventional methods. Electrophoresis 2001;22:187-207. DOI: 10.1002/1522-2683(200101)22:2%3C187::AID-ELPS187%3E3.0.CO;2-0

[4] Trentler TJ, Hickman KM, Goel SC, Viano AM, Gibbons PC, Buhro WE. Solution-liquid-solid growth of crystalline III-V semiconductors: An analogy to vapor-liquid-solid growth. Science 1995;270:1791-4. DOI: 10.1126/science.270.5243.1791

[5] Yang P, Wu Y, Fan R. Inorganic semiconductor nanowires. Int J Nanosci 2002;1:1-39. DOI: 10.1142/S0219581X02000061

[6] Hulteen J. A general template-based method for the preparation of nanomaterials. J Mater Chem 1997;7:1075-87. DOI: 10.1039/A700027H

[7] Zhao Y, Ye D, Wang G-C, Lu T-M. Designing nanostructures by glancing angle deposition. Proc.SPIE 5219. Nanotube and Nanowires 2003. DOI: 10.1117/12.505253

[8] Zhou CM, Gall D. Growth competition during glancing angle deposition of nanorod honeycomb arrays. Appl Phys Lett 2007;90:1-4. DOI: 10.1063/1.2709929

[9] Taschuk MT, Hawkeye MM, Brett MJ. Glancing Angle Deposition. Third Edit. Elsevier Ltd.; 2010. https://doi.org/10.1016/B978-0-8155-2031-3.00013-2.

[10] Kashyap KK, Choudhuri B, Chinnamuthu P. Enhanced Optical and Electrical Properties of Metallic Surface Plasmon Sensitized TiO_2 Nanowires. IEEE Trans Nanotechnol 2020;19:519-26. DOI: 10.1109/TNANO.2020.3004876

[11] Ngangbam C, Singh NK, Mondal A. Effect of Ag Doping on the Glancing Angle Deposition Synthesized TiO_2 Nanowire for Enlarged Photodetection . J Nanosci Nanotechnol 2017;18:5059-62. DOI: 10.1166/jnn.2018.15337

[12] Chinnamuthu P, Mondal A, Dhar JC, Singh NK. Visible light detection using glancing angle deposited TiO_2 nanowire arrays. Jpn J Appl Phys 2015;54:4-7. DOI: 10.7567/JJAP.54.06FJ01

[13] Dhar JC, Mondal A, Singh NK, Chinnamuthu P. Low-leakage TiO_2 nanowire dielectric MOS device using Ag schottky gate contact. IEEE Trans Nanotechnol 2013;12:948-50. DOI: 10.1109/TNANO.2013.2277600

[14] Mondal A, Dhar JC, Chinnamuthu P, Singh NK, Chattopadhyay KK, Das SK, et al. Electrical properties of vertically oriented TiO_2 nanowire arrays synthesized by glancing angle deposition technique. Electron Mater Lett 2013;9:213-7. DOI: 10.1007/s13391-012-2136-5

[15] Chinnamuthu P, Dhar JC, Mondal A, Bhattacharyya A, Singh NK. Ultraviolet detection using TiO_2 nanowire array with Ag Schottky contact. J Phys D Appl Phys 2012;45. DOI: 10.1088/0022-3727/45/13/135102

[16] Pooja P, Chinnamuthu P. Surface state controlled superior photodetection properties of isotype n-TiO_2/In_2O_3 heterostructure nanowire array with

high specific detectivity. IEEE Trans Nanotechnol 2020;19:34-41. DOI: 10.1109/TNANO.2019.2956960

[17] Chetri P, Dhar JC. Improved photodetector performance of SnO_2 nanowire by optimized air annealing. Semicond Sci Technol 2020;35:45014. DOI: 10.1088/1361-6641/ab7434

[18] Rajkumari R, Singh NK. Effect of Annealing on the Structural and Electrical Properties of GLAD Synthesized Vertical Aligned WO_3 Nanowire. IEEE Trans Nanotechnol 2019;18:676-83. DOI: 10.1109/TNANO.2019.2927367

[19] Panigrahy S, Dhar JC. Non-volatile memory property of Er_2O_3 doped SnO_2 nanowires synthesized using GLAD technique. J Mater Sci Mater Electron 2019;0:0. DOI: 10.1007/s10854-019-01151-0

[20] Panigrahy S, Dhar JC. Post Annealing Effects on Er_2O_3 Nanowire Arrays for Improved Photodetection. IEEE Trans Nanotechnol 2018;17:1189-96. DOI: 10.1109/TNANO.2018.2869223

[21] Lahiri R, Ghosh A, Choudhuri B, Mondal A. Investigation on improved performance of Erbium doped TiO_2 nanowire based UV detector. Mater Res Bull 2018;103:259-67. DOI: 10.1016/j.materresbull.2018.03.024

[22] Lahiri R, Mondal *A. superior* memory of Er-doped TiO_2 nanowire MOS capacitor. IEEE Electron Device Lett 2018;39:1856-9. DOI: 10.1109/LED.2018.2874272

[23] Choudhuri B, Mondal A, Dwivedi SMMD, Henini M. Fabrication of novel transparent Co_3O_4-TiO_2 nanowires p-n heterojunction diodes for multiband photodetection applications. J Alloys Compd 2017;712:7-14. DOI: 10.1016/j.jallcom.2017.04.068

[24] Singh NK, Mondal A, Dhar JC, Chakrabartty S, Chattopadhyay KK, Bhattacharyya A. Improved photo-detection from annealed SiO_x-$In_{2-x}O_{3-y}$ axial heterostructure nanocolumns. J Phys D Appl Phys 2014;47. DOI: 10.1088/0022-3727/47/10/105106

[25] Singh HM, Choudhuri B, Chinnamuthu P. Investigation of optoelectronic properties in germanium nanowire integrated silicon substrate using kelvin probe force microscopy. IEEE Trans Nanotechnol 2020;19;628-34. DOI: 10.1109/TNANO.2020.3010691

[26] Monroy E, Omnès F, Calle F. Wide-bandgap semiconductor ultraviolet photodetectors. Semicond Sci Technol 2003;18:R33. DOI: 10.1088/0268-1242/18/4/201

[27] Dorrer C, Rühe J. Some thoughts on superhydrophobic wetting. Soft Matter 2009;5:51-61. DOI: 10.1039/B811945G

[28] Byun D, Hong J, Ko JH, Lee YJ, Park HC, Byun B-K, et al. Wetting characteristics of insect wing surfaces. J Bionic Eng 2009;6:63-70. DOI: 10.1016/S1672-6529(08)60092-X

[29] Young T. III. An essay on the cohesion of fluids. Philos Trans R Soc London 1805:65-87. DOI: 10.1098/rstl.1805.0005

[30] Wenzel RN. Resistance of solid surfaces to wetting by water. Ind Eng Chem 1936;28:988-94. DOI: 10.1021/ie50320a024

[31] Cassie ABD, Baxter S. Wettability of porous surfaces. Trans Faraday Soc 1944;40:546-51. DOI: 10.1039/TF9444000546

[32] Miyauchi M, Nakajima A, Watanabe T, Hashimoto K. Photocatalysis and photoinduced hydrophilicity of various metal oxide thin films. Chem Mater 2002;14:2812-6. DOI: 10.1021/cm020076p

[33] Kenanakis G, Vernardou D, Katsarakis N. Light-induced self-cleaning properties of ZnO nanowires grown at low temperatures. Appl Catal A Gen 2012;411:7-14. DOI: 10.1016/j.apcata.2011.09.041

[34] Fujishima A, Zhang X, Tryk DA. TiO_2 photocatalysis and related surface phenomena. Surf Sci Rep 2008;63:515-82. DOI: 10.1016/j.surfrep.2008.10.001

[35] Yasui M, Katagiri K, Yamanaka S, Inumaru K. Molecular selective photocatalytic decomposition of alkylanilines by crystalline TiO_2 particles and their nanocomposites with mesoporous silica. RSC Adv 2012;2:11132-7. DOI: 10.1039/C2RA21009F

[36] Xu C, Yuan Y, Yuan R, Fu X. Enhanced photocatalytic performances of TiO_2-graphene hybrids on nitro-aromatics reduction to amino-aromatics. RSC Adv 2013;3:18002-8. DOI: 10.1039/C3RA42579G

[37] Pooja P, Choudhuri B, Saranyan V, Chinnamuthu P. Synthesis of coaxial TiO_2/In_2O_3 nanowire assembly using glancing angle deposition for wettability application. Appl Nanosci 2019;9:529-37. DOI: 10.1007/s13204-018-0936-0

[38] Talinungsang, Purkayastha DD, Krishna MG. Dopant controlled photoinduced hydrophilicity and photocatalytic activity of SnO_2 thin films. Appl Surf Sci 2018;447:724-31. DOI: 10.1016/j.apsusc.2018.04.028

[39] Miyauchi M, Kieda N, Hishita S, Mitsuhashi T, Nakajima A, Watanabe T, et al. Reversible wettability control of TiO_2 surface by light irradiation. Surf Sci 2002;511:401-7. DOI: 10.1016/S0039-6028(02)01551-0

[40] Xu Z, Ao Z, Chu D, Younis A, Li CM, Li S. Reversible hydrophobic to hydrophilic transition in graphene via water splitting induced by UV irradiation. Sci Rep 2014;4:1-9. DOI: 10.1038/srep06450

[41] Toh AGG, Nolan MG, Cai R, Butler DL. Reversible wetting of titanium dioxide films. Device Process Technol. Microelectron. MEMS, Photonics, Nanotechnol. IV, vol. 6800, International Society for Optics and Photonics; 2008, p. 680004. DOI: 10.1117/12.759438

[42] Pooja P, Chinnamuthu P. Annealing Effect of Glancing Angle Electron Beam Deposited TiO_2/In_2O_3 Nanowires Array on Surface Wettability. Sci Rep 2020;10:1-8. DOI: 10.1038/s41598-020-66150-2

Section 2

Group III-V Compounds

Chapter 4

Recent Progress in Gallium Nitride for Photoelectrochemical Water Splitting

Fangliang Gao, Qing Liu, Jiang Shi and Shuti Li

Abstract

With the constant consumption of traditional energy sources, it is urgent to explore and develop new energy sources. Photoelectrochemical (PEC) water splitting is a method of preparing energy that can continuously generate hydrogen fuel without pollution to the environment. As an important part of the PEC water splitting system, the choice of semiconductor photoelectrode is crucial. Among these materials, gallium nitride (GaN) has attracted considerable attention due to its tunable band gap, favorable band edge positions, wide band gap, and good stability. In the past years, many reports have been obtained in GaN for PEC water splitting. This review summarizes the GaN as photoelectrodes for PEC water splitting, and methods to improve the efficiency of GaN for PEC water splitting also will be summarized from change morphology, doping, surface modification, and composition of solid solution or multiple-metal incorporation. Eventually, the future research directions and challenges of GaN for PEC water splitting are also discussed.

Keywords: photoelectrochemical water splitting, GaN, semiconductor

1. Introduction

Rising energy demand due to population growth has led to the rapid consumption of fossil fuels and serious environmental problems [1]. Currently, most of the world's energy comes from fossil fuels, which will eventually lead to its predictable depletion. The decline of fossil energy reserves and the urgency to reduce greenhouse gas emissions to alleviate climate warming is forcing us to seek a cleaner, more renewable, and sustainable alternative energy source [2, 3]. Hydrogen is considered as a future ideal energy carrier to replace fossil fuels due to its high gravimetric energy density and zero carbon emissions [4–6]. But the achievement of this clean energy scheme largely depends on economically efficient hydrogen production technologies. At present, the industrial production of hydrogen is mainly realized by the reforming of hydrocarbon steam in fossil energy or coal through reaction to fossil fuels under the control of steam, which is not only expensive but also causes large emissions [7, 8]. Therefore, the use of renewable energy to produce hydrogen is considered, despite challenges stand in the way [7, 9].

In recent years, solar energy has attracted much attention as the largest renewable energy source on the planet. If solar energy can be effectively used, it will

 IntechOpen

provide a continuous supply of energy for future energy [10, 11]. However, the vision of solar power to provide a significant portion of the global infrastructure is far from being realized. The main challenge comes from not having a cost-effective way to store solar energy. Solar water splitting is a prospective, environmentally friendly, and sustainable method to achieve this beautiful vision [10, 12, 13]. There are three types of solar water decomposition systems, photovoltaic electrolysis (PV-E), photochemical (PC) systems, and photoelectrochemical (PEC) cells, as shown in **Figure 1**. PV-E is achieved by connecting the photovoltaic cell and water electrolyzer. The advantage of this strategy is its solar-hydrogen conversion efficiency of more than 10%, but it is still too expensive compared to traditional hydrogen production methods [14–18]. The maturity of PV-E technology also determines that it is difficult to improve efficiency, so it is particularly important to find economical and suitable solar-hydrogen conversion methods. PC is a simple and cost-effective solar-hydrogen conversion method, but its conversion efficiency is less than 1%. In addition, the potentially explosive hydrogen-oxygen mixture produced requires expensive equipment for separation to avoid reaction, which greatly increases production costs [19]. In this case, PEC provides considerable conversion efficiency at an affordable cost [20, 21]. PEC integrates the light absorption and electrochemical processes of PV-E into a single unit. Two gases generated separately at the anode and cathode avoid further separation, which is helpful for reducing costs. If the conversion efficiency can reach 10% and the life span reaches 5 years, PEC is expected to be a replacement for traditional hydrogen production methods [22–24].

Basically, solar energy is converted into chemical energy stored in the form of hydrogen molecules by PEC devices [25, 26]. And a PEC device usually includes a metal electrode and a semiconductor photoelectrode. Ideally, semiconductors need to have a proper band gap and band structure to provide sufficient reaction potential and cover the solar spectrum as much as possible. In addition, excellent carrier transport performance and good physicochemical stability are also essential. Although a large number of semiconductor materials such as ZnO [27, 28], TiO_2 [29, 30], WO_3 [31, 32], and $BiVO_4$ [33, 34] have been studied for photohydrolysis experiments, no dependent material meets all the critical conditions described above. Usually, overall water splitting consists of two half-reactions: oxidation of water and reduction of protons.

$$2H_2O + 4h^+ \rightarrow 4H^+ + O_2\ E_{ox} = +1.23\ V \tag{1}$$

$$4H^+ + 4e^- \rightarrow 2H_2\quad E_{red} = 0\ V \tag{2}$$

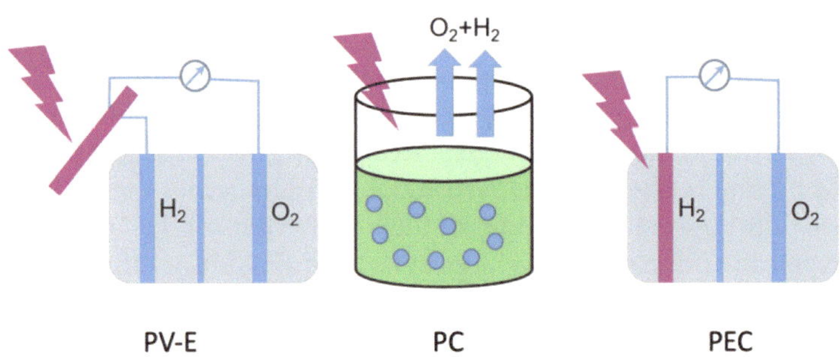

Figure 1.
Schematic of three types of solar water splitting system: PV-E, PC system, and PEC cell.

It can be seen from the equation that the minimum voltage for water splitting is 1.23 V, which requires that the energy absorbed by exciting an electron is not less than 1.23 eV. In order to meet this requirement, the photon energy absorbed by the photoelectrode must also be at least 1.23 eV. But in fact, the energy required due to the energy loss caused by the failure to reach the ideal structure is far more than 1.23 eV [35, 36].

In general, PEC water splitting includes the following processes:

1. Under light irradiation, carriers are generated in the semiconductor with a suitable band gap.

2. Photogenerated carriers separate and migrate to the surface of the semiconductor.

3. Redox reactions are induced by photogenerated carriers on semiconductor surfaces (**Figure 2**) [37, 38].

The number of photogenerated carriers is determined by the absorption efficiency of the semiconductor, which also reflects the utilization of sunlight. The separation and migration processes of carriers are related to how many can reach the semiconductor surface. Unfortunately, some carriers are lost resulting from recombining on their way to the surface. And the carriers that reach the surface of the semiconductor want to trigger an efficient water splitting reaction, which must meet the following requirements. First, the conduction band edge potential of the semiconductor material should be lower than H_2 evolution potential, while the valence band edge potential should be higher than O_2 evolution potential [39]. This means that the band gap of the semiconductor should be greater than 1.23 eV. Semiconductor materials need to have stronger absorption in the solar spectrum to generate more photogenerated carriers. Although wide band gap semiconductor materials are likely to meet matching at the band edge positions, the absorption of sunlight is very limited [7, 9, 40]. Second, carriers need to be separated and transmitted quickly to reduce recombination, thereby improving the utilization efficiency of photogenerated carriers for PEC water splitting. Finally, materials used for PEC water splitting should be cost-effective and have good stability in the catalytic process [41].

Among semiconductor materials commonly used in PEC water splitting, gallium nitride (GaN) has been regarded as a promising candidate [42, 43]. GaN is likely to

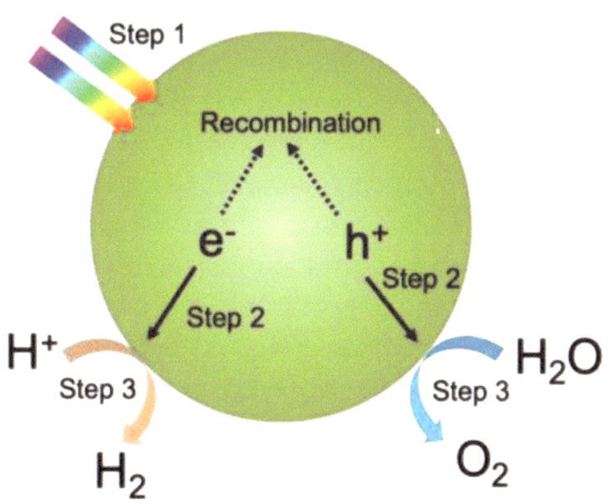

Figure 2.
Schematic illustration of typical PEC water splitting.

achieve self-driven overall water splitting because its band gap has good energy alignment with the water redox potential [43, 44]. In addition, GaN is inherently chemically inert even in a harsh environment, which guarantees the stability of the device [45, 46]. Furthermore, the band gap of GaN and its alloys can be tuned by alloying with Indium (In) to span nearly the entire solar spectrum [47, 48]. However, to achieve practical hydrogen production, GaN is still facing many challenges as an excellent photoelectrode material, including how to get a larger reaction area, how to enhance the absorption of light, and how to separate and transport photogenerated carriers more quickly and effectively [49, 50]. Correspondingly, many strategies have been proposed to address the mentioned drawbacks of GaN photoelectrode. Compared with thin-film and bulk counterpart, nanostructures have a smaller size and a larger surface area, which is helpful for shortening the transmission distance and promotes the separation of carriers. Thus the efficiency of carrier collection and utilization will be higher [51–53]. Doping is also one of the commonly used approaches to effectively improve the electrical and optical properties of GaN, which can directly tune the energy band structure and carrier transmission [54, 55]. Moreover, PEC water splitting kinetics can be promoted through the surface decoration of co-catalysts, which can enhance the transmission of carriers for water redox reaction [56, 57].

In this review, we summarize the recent progress of using GaN as photoelectrode for PEC water splitting and enumerate some commonly used strategies to improve the performance of photoelectrode. In the end, we also have a brief outlook of GaN for PEC water splitting.

2. Basic principles of solar water splitting

2.1 PEC cell configurations

In the introduction section, we briefly introduced the three types of solar water splitting. In this section, we will focus on the different structures of the PEC cell, which can be achieved by an n-type semiconductor as photoanode (or p-type semiconductor as photocathode) or connecting two different semiconductors.

For a semiconductor PEC cell with a half-reaction to occur on working electrode, a counter electrode is required to complete the other half-reaction circuit. Generally, a reference electrode is connected to the working electrode to characterize an externally applied voltage. If necessary, there are two compartments or ion exchange membranes between the working and counter electrodes to avoid product crossover. To overcome the thermodynamic obstacles of water splitting and the potential losses caused by the recombination process, the band gap of the working electrode is at least 1.6 eV [58–60]. However, the visible light absorption efficiency will be attenuated if the band gap is too wide. To solve this problem, that is, potential loss mechanisms that include reverse contact and overpotential caused by poor catalytic activity, the semiconductor material should be deposited on a highly conductive substrate to form a good ohmic contact, which allows most carriers to be quickly injected from the working electrode into the counter electrode [61, 62].

Obtaining enough photovoltage from a single photoelectrode to achieve solar water splitting is a challenge. It will be more favorable that combinates with dual semiconductors, because the second photoelectrode can replace the opposite electrode where the other half-reaction occurs and compensate for the lack of photovoltage [7]. To increase the light utilization, lighting should be irradiated from a larger band gap photoelectrode (transparent substrate) to a smaller band gap photoelectrode. In addition, these two semiconductors can form wireless back-to-back ohmic contacts, sharing a transparent conductive substrate [63]. By doing so, the potential loss in the electrolyte and the pH gradient between the two photoelectrodes can be reduced. Similarly, lighting should pass from a larger band gap material to a smaller band gap material. This series of battery structure is a relatively effective device [64].

2.2 Calculation of efficiencies

Comparing onset potentials and photocurrent density (normalized to the projected surface area of the photoelectrode) at 1.23 V versus RHE (photoanode) and 0 V versus RHE (photocathode) is a well-known method to evaluate the performance of water splitting. Since the product of water splitting is hydrogen, solar-to-hydrogen (STH) is the most critical parameter of merit to evaluate the performance and the efficiency of PEC water splitting on the device. It is defined as the following equation: [65].

$$\eta_{STH} = \left[\frac{\Phi_{H_2}(mols^{-1}m^{-2}) \times G^0_{f,H_2}(kJmol^{-1})}{P_{light}(Wm^{-2})}\right]_{AM\ 1.5G} \tag{3}$$

where Φ_{H2} is the hydrogen gas production rate, G^0_{f,H_2} is the Gibbs free energy of hydrogen gas (237 kJ mol^{-1} at 25°C), and P_{light} is the total solar irradiation input. The light source should match the solar spectrum of air mass 1.5 global (AM1.5 G). Since the redox reaction needs to consider the current loss, the Faraday efficiency needs to be considered. So, the STH formula is expressed as:

$$\eta_{STH} = \left[\frac{J_{SC}(mAcm^{-2}) \times 1.23\ V \times \eta_F}{P_{light}(mWm^{-2})}\right]_{AM\ 1.5G} \tag{4}$$

In general, J_{SC} can use current density instead of under zero bias and stable-state conditions. Applied bias photon to current conversion efficiency (ABPE) is also an important parameter for PEC water splitting systems, which is often used to evaluate the performance of a single photoelectrode independently. It can be written as: [65].

$$\text{ABPE} = \left[\frac{J_{SC}(mAcm^{-2})(\times 1.23\ V - V_{app}) \times \eta_F}{P_{light}(mWm^{-2})}\right]_{AM\ 1.5G} \tag{5}$$

where V_{app} is the applied potential between photoelectrode and the counter electrode.

It is important to understand the efficiency of photons to convert electrons/holes at certain wavelengths of PEC water splitting. Therefore, the incident photon-to-current conversion efficiency (IPCE) or external quantum efficiency (EQE) is proposed and expressed as: [65].

$$\text{IPCE}(\lambda) = \text{EQE}(\lambda) = \frac{electronflux\ (mols^{-1})}{photonflux\ (mols^{-1})} = \frac{|j_{ph}(mAcm^{-2})| \times hc\ (Vm)}{P_\lambda(mWcm^{-2}) \times \lambda(nm)} \tag{6}$$

where λ is the wavelength, P_λ is the incident light power, h is Planck's constant, c is the speed of light, and j_{ph} is the photocurrent density. Besides, integrating the IPCE value with the standard AM1.5G solar spectrum can estimate the total photocurrent density under solar light illumination. Its formula is defined as: [65].

$$J_{AM\,1.5} = \int (IPCE_\lambda \times \Phi_\lambda \times e)d\lambda \tag{7}$$

where e is the elementary electron and Φ_λ is photon flux of irradiation.

3. Approaches to increase GaN-based PEC water splitting efficiency

3.1 Surface decoration

Up to now, considerable efforts have been investigated on surface decoration to enhance PEC water splitting performance [66]. In this regard, various co-catalysts were studied by depositing on the surface of GaN to improve the efficiency of PEC water splitting. For instance, the quantum efficiency of the solid solution of GaN and ZnO for overall water splitting in the visible light region achieves the highest value of 2–3% after modified with a mixed oxide of Rh and Cr nanoparticles [67]. A Co-Pi catalyst photoelectron deposited on GaN thin-film photoelectrodes eliminated the anomalous two-plateau behavior and current spikes, which revealed that the Co-Pi catalyst is helpful for suppressing surface recombination and increases the photocurrent [47]. A similar but deeper achievement was carried out by Tricoli et al. for hybridizing highly transparent Co_3O_4 nano-island catalysts on GaN nanowire to enhance the water oxidation activity. The result shows that the per-metal turnover frequencies in 1 M NaOH aqueous solution are 0.34–0.65 s^{-1} at an overpotential of 400 mV, which is the best result of Co-based electrocatalysts until this report. This was attributed to Co_3O_4 that can play a role as hole scavenger, collecting photogenerated holes rapidly and suppressing carrier recombination [68]. Additionally, a size-controlled effect of poly-protected Rh nanoparticles on the photocatalytic activity of $(Ga_{1-x}Zn_x)(N_{1-x}O_x)$ was studied by Teranishi et al. for the first time. Their results show that the activity of smaller Rh cores is higher than the larger ones, which benefits from its increased surface area and improves charge separation efficiency [69]. This study was inconsistent with the previous report by Kamat et al. The greater the shift in the Fermi level observed in smaller gold nanoparticles, which is reflected in the higher photocatalytic reduction efficiency, the stronger the photocurrent [70].

Apart from nanoparticles, core-shell heterostructure is another important approach for surface decoration. GaN-InGaN core-shell rod arrays as photoanode for visible light-driven water splitting were studied by Waag et al. The core-shell structure extends the use of sunlight to the visible light region, thereby greatly improving the efficiency of water splitting. The photocurrent density of (0.3 mA/cm^2 at 1.35 V) GaN-InGaN was 10-fold higher than that of GaN (0.03 mA/cm^2 at 1.35 V), as shown in **Figure 3** [71]. Mi et al. employed GaN-InGaN core-shell nanowire for PEC water splitting, and the high incident photon-to-current conversion efficiency of up to ~27% is obtained [72]. It is expected to achieve higher PEC activity by surface treatment of GaN. And as far as the current development is concerned, it is foreseeable that surface modification is still a good strategy to achieve efficient water splitting.

3.2 GaN material having different morphologies

As an important part of the PEC water splitting system, the morphology of semiconductor materials is very important. Different morphologies have a great influence on the efficiency of PEC water splitting. Many different morphologies of GaN for PEC water splitting have been proposed. Xi and co-workers used

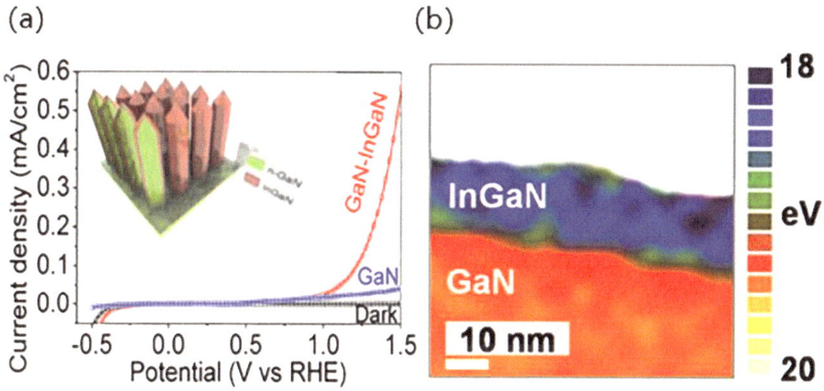

Figure 3.
(a) Current density of 3D GaN-InGaN core-shell rod array (red) and 3D GaN rod array (blue) in 0.01 M H_2SO_4 solution under 100 mW/cm^2 illumination using AM 1.5 filter [71]. (b) Plasmon energy map in the highlighted region showing the indium incorporation in the InGaN shell [71]. (a and b) Reproduced from Ref. [71] with permission from the American Chemical Society.

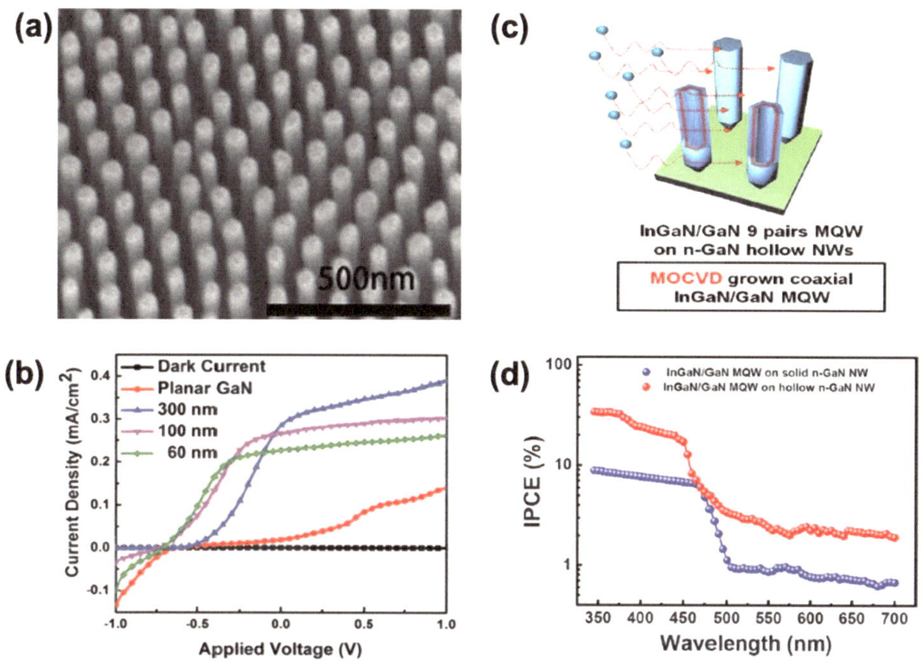

Figure 4.
(a) SEM images of GaN nanowires, (b) linear sweep voltammetry of GaN nanowires with diameters of 60, 100, and 300 nm and planar GaN [73]. (c) Structure schematic of InGaN/GaN MQW on n-GaN hollow NWs, (d) comparison of IPCE values for InGaN/GaN MQWs grown on solid and hollow n-GaN nanowires [74]. (a and b) Reproduced from Ref. [73] with permission from The Royal Society of Chemistry. (c and d) Reproduced from Ref. [74] with permission from Springer Nature.

metal–organic chemical vapor deposition (MOCVD) to fabricate GaN nanowires, and it has obtained high photocurrent density value at an applied bias voltage from −1 to 1 V [73]. Its morphology was shown in **Figure 4a**. It can be found from **Figure 4b** that compared to the planar structure and other diameters, 300 nm has a stronger current density due to a larger body-to-surface ratio, thereby increasing the efficiency of PEC water splitting. GaN microwires still have problems such as low crystal quality and light absorption. To further improve the efficiency of PEC water splitting, Park et al. used the plasma-assisted molecular beam epitaxy (PAMBE) technique to grow InGaN/GaN multiple quantum wells (MQWs) grown on hollow n-GaN nanowires (**Figure 4c**) [74]. The hollow and InGaN/GaN multiple quantum well structures of the nanowires allow the incident light to be refracted multiple times, increasing the absorption of light. **Figure 4d** shows the incident photon-to-current conversion efficiency value of the device, which can be found that the highest IPCE value of the device is as high as 33.3% and 415 μmol of hydrogen gas was generated within 1 hour.

Nanopores, nanocones, and honeycombs are other nanostructures of GaN. **Figure 5a** shows the GaN nanopore structure [43], nanopore structure used electrochemical lateral etching and ICP etching to prepare laterally porous, vertically holes well-ordered GaN. This structure reduces the UV reflectivity. The ordered vertical holes not only help open the embedded channels to the electrolyte on both sides and reduce the migration distance of bubbles in the water splitting reaction but also help to modulate the light field. Incident light can be modulated and captured into the nanopore to enhance the absorption of light, so the saturation photocurrent was 4.5 times that of the planar structure, as shown in **Figure 5d**. Moreover, GaN with aligned nanopore structure had been fabricated by combining MOCVD using a lateral anodic etching, as shown in **Figure 5b** [75]. Laterally porous 3D hierarchical nanostructures not only provided a large contact area between the electrode and the electrolyte but also increased the absorption of light and provided a channel for the transmission of light and electrons. The device also achieved high values of photocurrent of 0.32 mA/cm^2 by using etching voltages at 10 V (**Figure 5e**). Kim et al. had prepared GaN truncated nanocones [76], which was shown in **Figure 5c**. GaN truncated nanocones have concentrated incident light inside the nanostructure and enhanced the light trapping with reduced light losses from surface reflection. The relationship between current density and potential was shown in **Figure 5f**, which indicated that the photocurrent of GaN truncated nanocones was three times higher than the planar structure.

The above structures are expected, and GaN can also have nanorods [77], nanocolumns [78], nano-pyramids [79], and so on. It can be known from the above results that changing the morphology of GaN influences the efficiency of PEC water

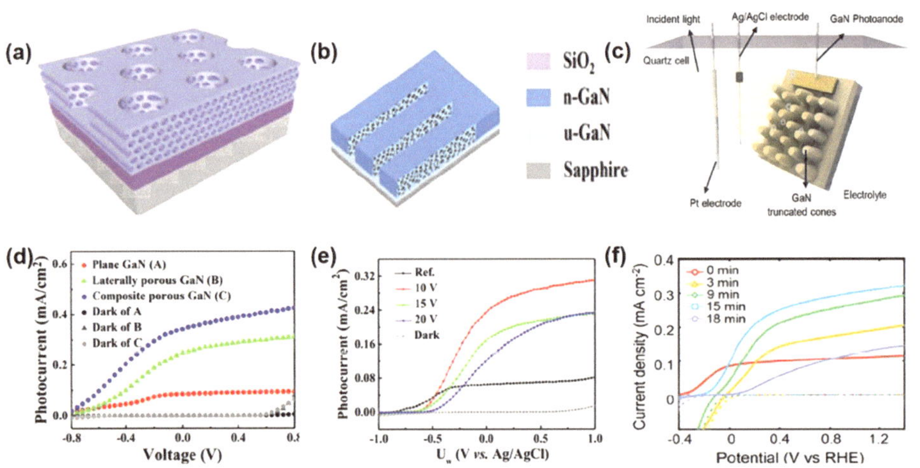

Figure 5.
(a) [43], (b) [75], and (c) [76] are the structure schematics of the composite porous GaN, laterally porous GaN, and photoelectrochemical cell, respectively. (d) Relation curves between photocurrent density and voltage of the above three GaN photoelectrodes [43]. (e) The photocurrent and applied current curves under different GaN etching voltages [75]. (f) Dark (segment) and illumination (straight) conditions for photocurrent density in linear scanning voltammetry [76]. (a and d), (b and e), and (c and f) reproduced from Ref. [43, 75, 76], respectively, with permission from the American Chemical Society.

splitting, which mainly affects the light absorption efficiency of GaN and reduces light reflection and loss. Therefore, it is very important to choose the appropriate semiconductor morphology for PEC water splitting system.

3.3 GaN material having different doping

Doping is a commonly used and effective method to improve the performance of materials. It mainly adjusts the energy band of the material, so that the photogenerated electrons and holes are better transported and high efficiency of PEC water splitting is obtained. Zhou and co-workers doped ZnO-GaN (GZNO) solid solution with La, as shown in **Figure 6a** [80]. La-dopant incorporation is optimized to adjust the bending of the band gap, which increases the thickness of the space charge region, thereby improving the separation of photogenerated carriers. **Figure 6c** shows the photocatalytic performance of GZNO and 3% La GZNO. It can be clearly seen that the photocatalyst doped with La produces more hydrogen and oxygen under the same conditions, which indicates that the performance of the photocatalyst is significantly improved after doping. **Figure 6b** shows the schematic of Ni-doped AlN and two-dimensional GaN monolayers [81]. By controlling the doping content of Ni, it can adjust the band bending of GaN. **Figure 6d** displays the binding strength of GaN and AlN composites with different transition metals doped. It can be found that Ni doping is the best for OER because they have small OER overpotentials.

GaN doped with Mn [82], Mg [83], or CrO are also reported [84]. Doping is also a good method to improve the efficiency of PEC water splitting. It mainly adjusts the energy band of GaN through doping, thereby promoting the separation of photogenerated electrons and holes and effectively preventing the recombination of

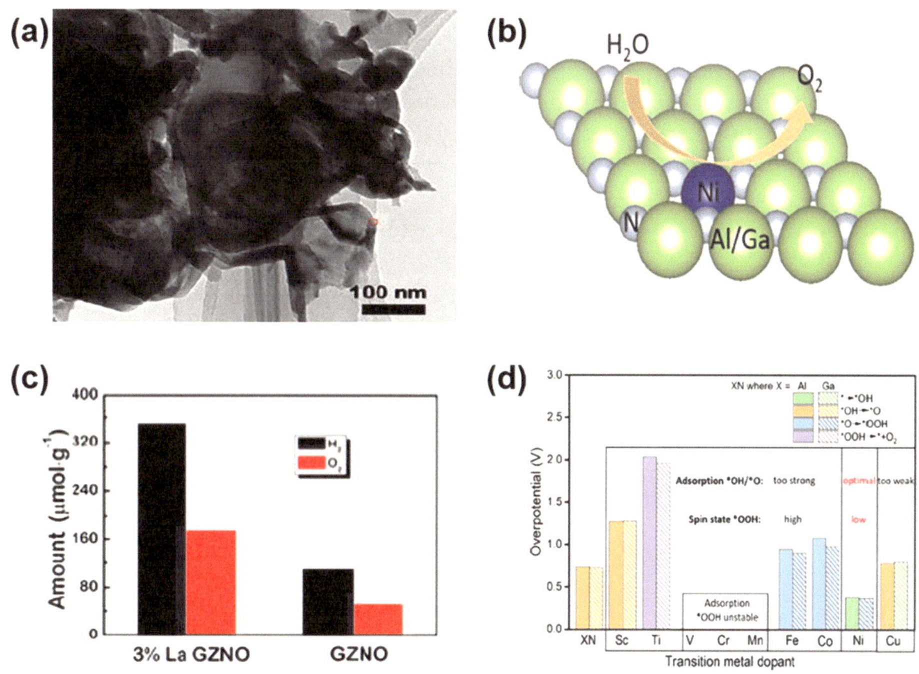

Figure 6.
(a) TEM images of 3% La GZNO [80]. (b) Schematic diagram of Ni-doped structure [81]. (c) The amount of H_2 and O_2 produced by overall water splitting after 8 hours of GZNO and 3% La GZNO [80]. (d) Binding strength of OH or O of different transition metal-doped GaN and AlN composites [81]. (a and c) Reproduced from Ref. [73] with permission from The Royal Society of Chemistry. (b and d) Reproduced from Ref. [74] with permission from the American Chemical Society.

carriers. However, excessive doping will deteriorate the crystal quality of the material. So, it is important to choose the doping material and control dopant incorporation.

3.4 Composition of solid solution

The solid solution is a wurtzite structure composite material composed of GaN and ZnO mixed in a certain proportion. It adjusts the doping content of ZnO to change the band gap of the solid solution and realizes PEC of water splitting under the visible light. This concept was first proposed by Maeda and co-workers [85]. And then, Ohno et al. used $Rh_{2-y}Cr_yO_3$ nanoparticles to modify the solid solution, and the device shows outstanding stability; it has been working continuously for half a year under light irradiation, as shown in **Figure 7a** [86]. The co-catalyst is beneficial to suppress the oxidative decomposition of the solid solution, thereby making the device more stable. NiCoFeP and flux-assisted method can also be used to modify the solid solution to improve the efficiency of PEC water splitting [87, 88]. The conversion efficiency of solar energy using NiCoFeP-modified solid solution exceeds 1% at 1.23 V vs. RHE. To further improve the efficiency of PEC water splitting, solid solution nanosheets modified with Rh nanoparticles have been proposed, as shown in **Figure 7d** [89]. This shows 0.7 μmol h^{-1} g^{-1} of hydrogen production in an aqueous H_2SO_4 solution. The nitridation process was used to change the morphology from hexagonal 2D $ZnGa_2O_4$ nanosheets to 2D $(GaN)_{1-x}(ZnO)_x$ nanosheets, reducing the path of carrier transportation and decreasing the recombination of electrons and holes. So, the composition of a solid solution or multiple-metal incorporation can expand the light absorption range of the device, improving the absorption of light and increasing the efficiency of PEC water splitting.

3.5 The multiple-metal incorporation

The method of forming multiple-metal incorporation is similar to that of a solid solution. Different In content incorporation can change the band gap of GaN to widen the absorption spectrum range. Many different multiple-metal incorporations have been proposed [90–94]. AlOtaibi et al. grown InGaN/GaN core-shell

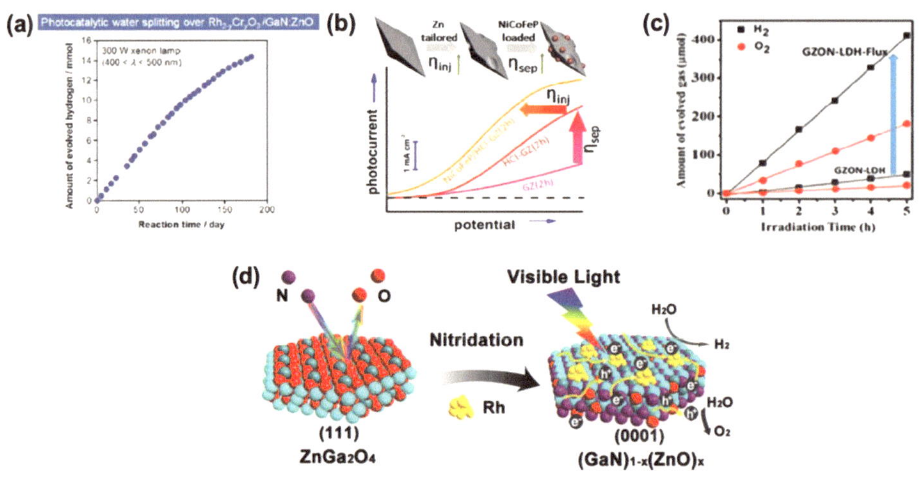

Figure 7.
(a) Curve of total hydrogen production over time [86]. (b) The relationship between photocurrent and voltage for GaN-ZnO in different ways of treatment [87]. (c) Variation curve of the amount of hydrogen and oxygen produced with time of $(GaN)_{1-x}(ZnO)_x$ solid solutions from Zn_2Ga-LDH modified with $Rh_{2-y}CryO_3$ nanoparticles [88]. (d) Structure conversion flowchart from 2D $ZnGa_2O_4$ to 2D $(GaN)_{1-x}(ZnO)_x$ [89]. (a–d) Reproduced from Ref. 86–89 with permission from the American Chemical Society.

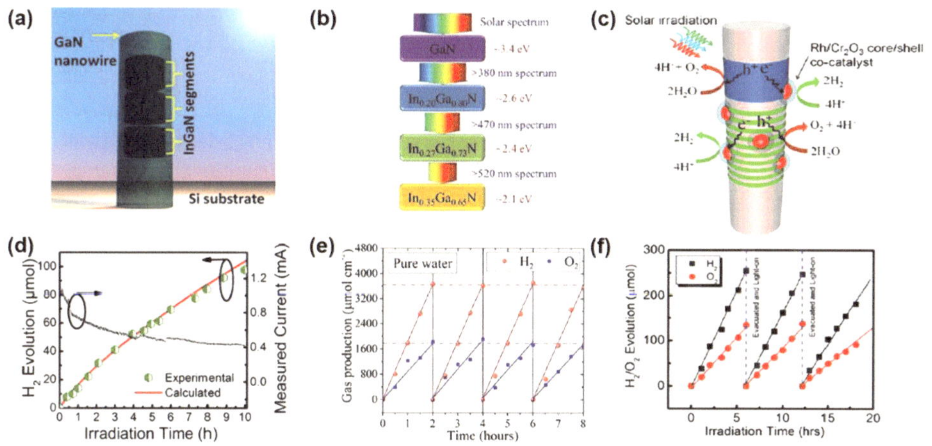

Figure 8.
(a) Schematic of the InGaN/GaN core-shell structure [72]. (b) Schematic diagram of ideal light absorption structure of a multiband InGaN stack with different indium compositions [95]. (c) Schematic diagram of photocatalytic overall water splitting reaction mechanism [48]. (d) Hydrogen production in 1 mol/L HBr at 0.2 V vs. the counter electrode [72]. (e) Hydrogen and oxygen produced as a function of time under multiple experiment cycles [95]. (f) Hydrogen and oxygen evolution as a function of irradiation time under full arc (>300 nm) 300 W xenon lamp irradiation [48]. (a and d) and (c and e) Reproduced from Ref. [48, 72] with permission from the American Chemical Society. (b and f) Reproduced from Ref. [95] with permission from The Royal Society of Chemistry.

nanowire arrays on Si substrate by catalyst-free MBE, as shown in **Figure 8a** [72]. It has a photoelectric conversion efficiency of up to 27% under ultraviolet and visible light irradiation. The photoelectrode continued to work for 10 hours, and the hydrogen production was consistent with the theoretical value (**Figure 8d**), which indicates that the photoelectrode has good stability and hydrogen production ability. And the quadruple-band InGaN nanowire arrays were integrated on a nonpolar substrate, which includes $In_{0.35}Ga_{0.65}N$, $In_{0.27}Ga_{0.73}N$, $In_{0.20}Ga_{0.80}N$, and GaN and exhibits a solar-to-hydrogen efficiency of ~5.2% in a relatively stable state (**Figure 8b**) [95]. Multiband nanowire arrays enhance light absorption to improve the performance of PEC water splitting. Moreover, the multiband nanowire array photoelectrode has good stability and high photocatalyst efficiency for overall water splitting, as shown in **Figure 8e**. To improve the efficiency of the photolysis of water, InGaN heterostructures have been proposed. Kibria and co-workers have fabricated InGaN/GaN nanowire heterostructures, in which the internal quantum efficiency is about 13% [48]. The nanowire heterostructure is shown in **Figure 8c**. The combination of GaN and InGaN expands the light absorption range of GaN from ultraviolet light to visible light, which greatly improves the light absorption range and improves the efficiency of photolysis. The InGaN/GaN nanowire heterostructure photoelectrode also exhibits extremely high stability and high hydrogen production capabilities, as shown in **Figure 8f**. Moreover, nanowire arrays [96], tunnel junction nanowire [97], have also been reported.

In summary, the multiple-metal incorporation can greatly improve the efficiency of PEC water splitting of GaN. The structures and In content will greatly affect the efficiency of PEC water splitting. So, it is important to choose a suitable structure and the In content while preparing the GaN-based photoelectrode.

4. Conclusion

This review mainly introduces the application of GaN in the PEC water splitting system and summarizes the methods to improve the efficiency of PEC water splitting. The methods to enhance efficiency are mainly carried out in the following four aspects, such as morphology, doping, surface modification, and composition of solid solution or multiple-metal incorporation. Up to now, GaN has made great progress in the application of PEC water splitting; the solar-to-hydrogen efficiency of 12.6% has already been obtained without any external bias [98], better than CoP catalyst electrodes (6.7%) reported recently [99], but it still not as excellent as TiO_2 (18.5%) [100]. And its properties need to be further optimized to improve the absorption efficiency of visible light, increase the carrier migration speed, and facilitate carrier transport. The follow-up works are suggested from the following aspects:

1. At present, most water splitting processes are carried out in alkaline or acidic solutions. It should be considered how to ensure the stability and catalytic activity of metal nitrides for a long time.

2. Although the theory of water splitting is simple, the reaction process is still not clear, and in-depth study of the mechanism is helpful for the design of the catalyst.

3. Reasonable design of the composition and structure of the catalyst to adjust its electronic structure, band gap, band edge potential, and microstructure help to improve the catalytic performance. We believe that with the deepening of research, the efficiency of GaN for water splitting can be further promoted.

Conflict of interest

The authors declare no conflict of interest.

Author details

Fangliang Gao*, Qing Liu, Jiang Shi and Shuti Li*
Guangdong Engineering Research Center of Optoelectronic Functional Materials and Devices, Institute of Semiconductors, South China Normal University, Guangzhou, China

*Address all correspondence to: gaofl@m.scnu.edu.cn; lishuti@scnu.edu.cn

References

[1] Lee R. The outlook for population growth. Science. 2011;**333**(6042): 569-573. DOI: 10.1126/science.1208859

[2] Di Primio R, Horsfield B, Guzman-Vega MA. Determining the temperature of petroleum formation from the kinetic properties of petroleum asphaltenes. Nature. 2000;**406**(6792):173-176. DOI: 10.1038/35018046

[3] Christopher K, Dimitrios R. A review on exergy comparison of hydrogen production methods from renewable energy sources. Energy & Environmental Science. 2012;**5**(5): 6640-6651. DOI: 10.1039/c2ee01098d

[4] Wang W, Xu X, Zhou W, et al. Recent progress in metal-organic frameworks for applications in electrocatalytic and photocatalytic water splitting. Advanced Science. 2017; **4**(4):1600371. DOI: 10.1002/advs.201600371

[5] Tamirat AG, Rick J, Dubale AA, et al. Using hematite for photoelectrochemical water splitting: A review of current progress and challenges. Nanoscale Horizons. 2016;**1**(4):243-267. DOI: 10.1039/C5NH00098J

[6] Linic S, Christopher P, Ingram DB. Plasmonic-metal nanostructures for efficient conversion of solar to chemical energy. Nature Materials. 2011;**10**(12): 911-921. DOI: 10.1038/nmat3151

[7] Tee SY, Win KY, Teo WS, et al. Recent progress in energy-driven water splitting. Advanced Science. 2017; **4**(5):1600337. DOI: 10.1002/advs.201600337

[8] Chiarello GL, Aguirre MH, Selli E. Hydrogen production by photocatalytic steam reforming of methanol on noble metal-modified TiO_2. Journal of Catalysis. 2010;**273**(2):182-190. DOI: 10.1016/j.jcat.2010.05.012

[9] Jafari T, Moharreri E, Amin AS, et al. Photocatalytic water splitting—The untamed dream: A review of recent advances. Molecules. 2016;**21**(7):900. DOI: 10.3390/molecules21070900

[10] Tachibana Y, Vayssieres L, Durrant JR. Artificial photosynthesis for solar water-splitting. Nature Photonics. 2012;**6**(8):511. DOI: 10.1038/nphoton.2012.175

[11] Joy J, Mathew J, George SC. Nanomaterials for photoelectrochemical water splitting—Review. International Journal of Hydrogen Energy. 2018; **43**(10):4804-4817. DOI: 10.1016/j.ijhydene.2018.01.099

[12] Chen X, Shen S, Guo L, et al. Semiconductor-based photocatalytic hydrogen generation. Chemical Reviews. 2010;**110**(11):6503-6570. DOI: 10.1021/cr1001645

[13] Mao SS, Shen S. Catalysing artificial photosynthesis. Nature Photonics. 2013; **7**(12):944-946. DOI: 10.1038/nphoton.2013.326

[14] Luo J, Im JH, Mayer MT, et al. Water photolysis at 12.3% efficiency via perovskite photovoltaics and earth-abundant catalysts. Science. 2014; **345**(6204):1593-1596. DOI: 10.1126/science.1258307

[15] Jia J, Seitz LC, Benck JD, et al. Solar water splitting by photovoltaic-electrolysis with a solar-to-hydrogen efficiency over 30%. Nature Communications. 2016;7(1):1-6. DOI: 10.1038/ncomms13237

[16] Cox CR, Lee JZ, Nocera DG, et al. Ten-percent solar-to-fuel conversion with nonprecious materials. Proceedings of the National Academy of Sciences. 2014;**111**(39):14057-14061. DOI: 10.1073/pnas.1414290111

[17] Peharz G, Dimroth F, Wittstadt U. Solar hydrogen production by water splitting with a conversion efficiency of 18%. International Journal of Hydrogen Energy. 2007;**32**(15):3248-3252. DOI: 10.1016/j.ijhydene.2007.04.036

[18] Khaselev O, Bansal A, Turner JA. High-efficiency integrated multijunction photovoltaic/electrolysis systems for hydrogen production. International Journal of Hydrogen Energy. 2001;**26**(2):127-132. DOI: 10.1016/s0360-3199(00)00039-2

[19] Fabian DM, Hu S, Singh N, et al. Particle suspension reactors and materials for solar-driven water splitting. Energy & Environmental Science. 2015;**8**(10):2825-2850. DOI: 10.1039/c5ee01434d

[20] Ager JW, Shaner MR, Walczak KA, et al. Experimental demonstrations of spontaneous, solar-driven photoelectrochemical water splitting. Energy & Environmental Science. 2015; **8**(10):2811-2824. DOI: 10.1039/c5ee00457h

[21] McKone JR, Lewis NS, Gray HB. Will solar-driven water-splitting devices see the light of day? Chemistry of Materials. 2014;**26**(1):407-414. DOI: 10.1021/cm4021518

[22] Sathre R, Scown CD, Morrow WR, et al. Life-cycle net energy assessment of large-scale hydrogen production via photoelectrochemical water splitting. Energy & Environmental Science. 2014; **7**(10):3264-3278. DOI: 10.1039/c4ee01019a

[23] Kment S, Schmuki P, Hubicka Z, et al. Photoanodes with fully controllable texture: The enhanced water splitting efficiency of thin hematite films exhibiting solely (110) crystal orientation. ACS Nano. 2015; **9**(7):7113-7123. DOI: 10.1021/acsnano.5b01740

[24] Ding C, Shi J, Wang Z, et al. Photoelectrocatalytic water splitting: Significance of cocatalysts, electrolyte, and interfaces. ACS Catalysis. 2017;**7**(1): 675-688. DOI: 10.1021/acscatal.6b03107

[25] Hensel J, Wang G, Li Y, et al. Synergistic effect of CdSe quantum dot sensitization and nitrogen doping of TiO_2 nanostructures for photoelectrochemical solar hydrogen generation. Nano Letters. 2010;**10**(2): 478-483. DOI: 10.1021/nl903217w

[26] Li W, Li J, Wang X, et al. Photoelectrochemical and physical properties of WO_3 films obtained by the polymeric precursor method. International Journal of Hydrogen Energy. 2010;**35**(24):13137-13145. DOI: 10.1016/j.ijhydene.2010.09.011

[27] Sun Y, Xu B, Shen Q, et al. Rapid and efficient self-assembly of Au@ ZnO core-shell nanoparticle arrays with an enhanced and tunable plasmonic absorption for photoelectrochemical hydrogen generation. ACS Applied Materials & Interfaces. 2017;**9**(37): 31897-31906. DOI: 10.1021/acsami.7b09325

[28] Wu M, Chen WJ, Shen YH, et al. In situ growth of matchlike ZnO/Au plasmonic heterostructure for enhanced photoelectrochemical water splitting. ACS Applied Materials & Interfaces. 2014;**6**(17):15052-15060. DOI: 10.1021/am503044f

[29] Cho IS, Logar M, Lee CH, et al. Rapid and controllable flame reduction of TiO_2 nanowires for enhanced solar water-splitting. Nano Letters. 2014; **14**(1):24-31. DOI: 10.1021/nl4026902

[30] Yang JS, Wu JJ. Toward eco-friendly and highly efficient solar water splitting using In_2S_3/anatase/rutile TiO_2 dual-staggered-heterojunction nanodendrite array photoanode. ACS Applied Materials & Interfaces. 2018;

10(4):3714-3722. DOI: 10.1021/acsami.7b19139

[31] Feng X, Chen Y, Qin Z, et al. Facile fabrication of sandwich structured WO_3 nanoplate arrays for efficient photoelectrochemical water splitting. ACS Applied Materials & Interfaces. 2016;**8**(28):18089-18096. DOI: 10.1021/acsami.6b04887

[32] Zheng G, Wang J, Zu G, et al. Sandwich structured WO_3 nanoplatelets for highly efficient photoelectrochemical water splitting. Journal of Materials Chemistry A. 2019;**7**(45):26077-26088. DOI: 10.1039/c9ta09188b

[33] Xue D, Kan M, Qian X, et al. A tandem water splitting cell based on nanoporous $BiVO_4$ photoanode cocatalyzed by ultrasmall cobalt borate sandwiched with conformal TiO_2 layers. ACS Sustainable Chemistry & Engineering. 2018;**6**(12):16228-16234. DOI: 10.1021/acssuschemeng.8b03078

[34] Kim JH, Jo Y, Kim JH, et al. Wireless solar water splitting device with robust cobalt-catalyzed, dual-doped $BiVO_4$ photoanode and perovskite solar cell in tandem: A dual absorber artificial leaf. ACS Nano. 2015;**9**(12):11820-11829. DOI: 10.1021/acsnano.5b03859

[35] Bak T, Nowotny J, Rekas M, et al. Photo-electrochemical hydrogen generation from water using solar energy. Materials-related aspects. International Journal of Hydrogen Energy. 2002;**27**(10):991-1022. DOI: 10.1016/s0360-3199(02)00022-8

[36] Werner JH, Kolodinski S, Queisser HJ. Novel optimization principles and efficiency limits for semiconductor solar cells. Physical Review Letters. 1994;**72**(24):3851. DOI: 10.1103/physrevlett.72.3851

[37] Wang D, Pierre A, Kibria MG, et al. Wafer-level photocatalytic water splitting on GaN nanowire arrays grown by molecular beam epitaxy. Nano Letters. 2011;**11**(6):2353-2357. DOI: 10.1021/nl2006802

[38] Kudo A, Miseki Y. Heterogeneous photocatalyst materials for water splitting. Chemical Society Reviews. 2009;**38**(1):253-278. DOI: 10.1039/b800489g

[39] Abe R. Recent progress on photocatalytic and photoelectrochemical water splitting under visible light irradiation. Journal of Photochemistry and Photobiology C: Photochemistry Reviews. 2010;**11**(4):179-209. DOI: 10.1039/c3cs60378d

[40] Turner JA. A realizable renewable energy future. Science. 1999;**285**(5428):687-689. DOI: 10.1126/science.285.5428.687

[41] Li Y, Zhang JZ. Hydrogen generation from photoelectrochemical water splitting based on nanomaterials. Laser & Photonics Reviews. 2010;**4**(4):517-528. DOI: 10.1002/lpor.200910025

[42] Zhang Z, Qian Q, Li B, et al. Interface engineering of monolayer MoS_2/GaN hybrid heterostructure: Modified band alignment for photocatalytic water splitting application by nitridation treatment. ACS Applied Materials & Interfaces. 2018;**10**(20):17419-17426. DOI: 10.1021/acsami.8b01286

[43] Yang C, Xi X, Yu Z, et al. Light modulation and water splitting enhancement using a composite porous GaN structure. ACS Applied Materials & Interfaces. 2018;**10**(6):5492-5497. DOI: 10.1021/acsami.7b15344

[44] Tseng WJ, van Dorp DH, Lieten RR, et al. Anodic etching of n-GaN epilayer into porous GaN and its photoelectrochemical properties. The Journal of Physical Chemistry C. 2014;

118(51):29492-29498. DOI: 10.1021/jp508314q

[45] Neumayer DA, Ekerdt JG. Growth of group III nitrides. A review of precursors and techniques. Chemistry of Materials. 1996;**8**(1):9-25. DOI: 10.1021/cm950108r

[46] Benton J, Bai J, Wang T. Utilisation of GaN and InGaN/GaN with nanoporous structures for water splitting. Applied Physics Letters. 2014; **105**(22):223902. DOI: 10.1063/1.4903246

[47] Kamimura J, Bogdanoff P, Abdi FF, et al. Photoelectrochemical properties of GaN photoanodes with cobalt phosphate catalyst for solar water splitting in neutral electrolyte. The Journal of Physical Chemistry C. 2017;**121**(23): 12540-12545. DOI: 10.1021/acs.jpcc.7b02253

[48] Kibria MG, Nguyen HPT, Cui K, et al. One-step overall water splitting under visible light using multiband InGaN/GaN nanowire heterostructures. ACS Nano. 2013;**7**(9):7886-7893. DOI: 10.1021/nn4028823

[49] Varadhan P, Fu HC, Priante D, et al. Surface passivation of GaN nanowires for enhanced photoelectrochemical water-splitting. Nano Letters. 2017; **17**(3):1520-1528. DOI: 10.1021/acs.nanolett.6b04559

[50] Walsh A, Yan Y, Huda MN, et al. Band edge electronic structure of $BiVO_4$: Elucidating the role of the Bi s and V d orbitals. Chemistry of Materials. 2009;**21**(3):547-551. DOI: 10.1021/cm802894z

[51] Tian J, Zhao Z, Kumar A, et al. Recent progress in design, synthesis, and applications of one-dimensional TiO_2 nanostructured surface heterostructures: A review. Chemical Society Reviews. 2014;**43**(20): 6920-6937. DOI: 10.1039/c4cs00180j

[52] Kang Q, Cao J, Zhang Y, et al. Reduced TiO_2 nanotube arrays for photoelectrochemical water splitting. Journal of Materials Chemistry A. 2013; **1**(18):5766-5774. DOI: 10.1039/c3ta10689f

[53] Wang H, Wang X, Luo X, et al. Optimization of all figure-of-merits in well-aligned GaN microwire array based Schottky UV photodetectors by Si doping. ACS Photonics. 2019;**6**(8): 1972-1980. DOI: 10.1021/acsphotonics.9b00363

[54] Tang YB, Bo XH, Xu J, et al. Tunable p-type conductivity and transport properties of AlN nanowires via Mg doping. ACS Nano. 2011;**5**(5):3591-3598. DOI: 10.1021/nn200963k

[55] Mitrofanov O, Manfra M, Weimann N. Impact of Si doping on radio frequency dispersion in unpassivated GaN/AlGaN/GaN high-electron-mobility transistors grown by plasma-assisted molecular-beam epitaxy. Applied Physics Letters. 2003;**82**(24): 4361-4363. DOI: 10.1063/1.1582373

[56] Zhong DK, Choi S, Gamelin DR. Near-complete suppression of surface recombination in solar photoelectrolysis by "Co-Pi" catalyst-modified W:$BiVO_4$. Journal of the American Chemical Society. 2011;**133**(45):18370-18377. DOI: 10.1021/ja207348x

[57] Hisatomi T, Maeda K, Takanabe K, et al. Aspects of the water splitting mechanism on $(Ga_{1-x}Zn_x)(N_{1-x}O_x)$ photocatalyst modified with $Rh_{2-y}Cr_yO_3$ cocatalyst. The Journal of Physical Chemistry C. 2009;**113**(51): 21458-21466. DOI: 10.1021/jp9079662

[58] Walter MG, Warren EL, McKone JR, et al. Solar water splitting cells. Chemical Reviews. 2010;**110**(11): 6446-6473. DOI: 10.1021/cr1002326

[59] Pinaud BA, Benck JD, Seitz LC, et al. Technical and economic feasibility of

centralized facilities for solar hydrogen production via photocatalysis and photoelectrochemistry. Energy & Environmental Science. 2013;**6**(7): 1983-2002. DOI: 10.1039/c3ee40831k

[60] Weber MF, Dignam MJ. Efficiency of splitting water with semiconducting photoelectrodes. Journal of the Electrochemical Society. 1984;**131**(6):1258-1265. DOI: 10.1149/1.2115797

[61] Takanabe K. Photocatalytic water splitting: Quantitative approaches toward photocatalyst by design. ACS Catalysis. 2017;**7**(11):8006-8022. DOI: 10.1021/acscatal.7b02662

[62] Stolarczyk JK, Bhattacharyya S, Polavarapu L, et al. Challenges and prospects in solar water splitting and CO_2 reduction with inorganic and hybrid nanostructures. ACS Catalysis. 2018;**8**(4):3602-3635. DOI: 10.1021/acscatal.8b00791

[63] Nozik AJ. Photochemical diodes. Applied Physics Letters. 1977;**30**(11): 567-569. DOI: 10.1063/1.89262

[64] Sivula K, Van De Krol R. Semiconducting materials for photoelectrochemical energy conversion. Nature Reviews Materials. 2016;**1**(2):15010. DOI: 10.1038/natrevmats.2015.10

[65] Chu S, Li W, Yan Y, et al. Roadmap on solar water splitting: Current status and future prospects. Nano Futures. 2017;**1**(2):022001. DOI: 10.1088/2399-1984/aa88a1

[66] Sivula K, Le Formal F, Grätzel M. Solar water splitting: Progress using hematite (α-Fe_2O_3) photoelectrodes. ChemSusChem. 2011;**4**(4):432-449. DOI: 10.1002/cssc.201000416

[67] Maeda K, Teramura K, Saito N, et al. Overall water splitting using (oxy) nitride photocatalysts. Pure and Applied Chemistry. 2006;**78**(12):2267-2276. DOI: 10.1351/pac200678122267

[68] Liu G, Karuturi SK, Simonov AN, et al. Robust sub-monolayers of Co_3O_4 nano-islands: A highly transparent morphology for efficient water oxidation catalysis. Advanced Energy Materials. 2016;**6**(15):1600697. DOI: 10.1002/aenm.201600697

[69] Ikeda T, Xiong A, Yoshinaga T, et al. Polyol synthesis of size-controlled Rh nanoparticles and their application to photocatalytic overall water splitting under visible light. The Journal of Physical Chemistry C. 2013;**117**(6): 2467-2473. DOI: 10.1021/jp305968u

[70] Subramanian V, Wolf EE, Kamat PV. Catalysis with TiO_2/gold nanocomposites. Effect of metal particle size on the Fermi level equilibration. Journal of the American Chemical Society. 2004;**126**(15):4943-4950. DOI: 10.1021/ja0315199

[71] Caccamo L et al. Band engineered epitaxial 3D GaN-InGaN core-shell rod arrays as an advanced photoanode for visible-light-driven water splitting. ACS Applied Materials & Interfaces. 2014; **6**(4):2235-2240. DOI: 10.1021/am4058937

[72] AlOtaibi B, Nguyen HPT, Zhao S, et al. Highly stable photoelectrochemical water splitting and hydrogen generation using a double-band InGaN/GaN core/shell nanowire photoanode. Nano Letters. 2013;**13**(9):4356-4361. DOI: 10.1021/nl402156e

[73] Xi X, Li J, Ma Z, et al. Enhanced water splitting performance of GaN nanowires fabricated using anode aluminum oxide templates. RSC Advances. 2019;**9**(26):14937-14943. DOI: 10.1039/c9ra01188a

[74] Park JH, Mandal A, Kang S, et al. Hydrogen generation using non-polar coaxial InGaN/GaN multiple quantum

well structure formed on hollow n-GaN nanowires. Scientific Reports. 2016; **6**(1):1-10. DOI: 10.1038/srep31996

[75] Yang C, Liu L, Zhu S, et al. GaN with laterally aligned nanopores to enhance the water splitting. The Journal of Physical Chemistry C. 2017;**121**(13): 7331-7336

[76] Kim YJ, Lee GJ, Kim S, et al. Efficient light absorption by GaN truncated nanocones for high performance water splitting applications. ACS Applied Materials & Interfaces. 2018;**10**(34):28672-28678. DOI: 10.1021/acsami.8b09084

[77] Tyagi P, Ramesh C, Kaswan J, et al. Direct growth of self-aligned single-crystalline GaN nanorod array on flexible Ta foil for photocatalytic solar water-splitting. Journal of Alloys and Compounds. 2019;**805**:97-103. DOI: 10.1016/j.jallcom.2019.07.071

[78] Xi X, Yang C, Cao H, et al. GaN nanocolumns fabricated by self-assembly Ni mask and its enhanced photocatalytic performance in water splitting. Applied Surface Science. 2018; **462**:310-315. DOI: 10.1016/j.apsusc.2018.08.113

[79] Hou Y, Yu X, Syed ZA, et al. GaN nano-pyramid arrays as an efficient photoelectrode for solar water splitting. Nanotechnology. 2016;**27**(45):455401. DOI: 10.1088/0957-4484/27/45/455401

[80] Zhou Y, Chen G, Yu Y, et al. Effects of La-doping on charge separation behavior of ZnO:GaN for its enhanced photocatalytic performance. Catalysis Science & Technology. 2016;**6**(4): 1033-1041. DOI: 10.1039/c5cy01193k

[81] Liang Q, Brocks G, Zhang X, et al. Monolayer nitrides doped with transition metals as efficient catalysts for water oxidation: The singular role of nickel. The Journal of Physical Chemistry C. 2019;**123**(43): 26289-26298. DOI: 10.1021/acs.jpcc.9b06704

[82] Liu SY, Sheu JK, Lin YC, et al. Mn-doped GaN as photoelectrodes for the photoelectrolysis of water under visible light. Optics Express. 2012;**20**(105): A678-A683. DOI: 10.1364/oe.20.00a678

[83] Kibria MG, Qiao R, Yang W, et al. Atomic-scale origin of long-term stability and high performance of p-GaN nanowire arrays for photocatalytic overall pure water splitting. Advanced Materials. 2016;**28**(38):8388-8397. DOI: 10.1002/adma.201602274

[84] Pan H, Gu B, Eres G, et al. Ab initio study on noncompensated CrO codoping of GaN for enhanced solar energy conversion. The Journal of Chemical Physics. 2010;**132**(10):104501. DOI: 10.1063/1.3337919

[85] Maeda K, Takata T, Hara M, et al. GaN:ZnO solid solution as a photocatalyst for visible-light-driven overall water splitting. Journal of the American Chemical Society. 2005; **127**(23):8286-8287. DOI: 10.1021/ja0518777

[86] Ohno T, Bai L, Hisatomi T, et al. Photocatalytic water splitting using modified GaN:ZnO solid solution under visible light: Long-time operation and regeneration of activity. Journal of the American Chemical Society. 2012; **134**(19):8254-8259. DOI: 10.1021/ja302479f

[87] Wang Z, Zong X, Gao Y, et al. Promoting charge separation and injection by optimizing the interfaces of GaN:ZnO photoanode for efficient solar water oxidation. ACS Applied Materials & Interfaces. 2017;**9**(36):30696-30702. DOI: 10.1021/acsami.7b09021

[88] Yang JH, Pei YR, Kim SJ, et al. Highly enhanced photocatalytic water-splitting activity of gallium zinc oxynitride derived from flux-assisted

Zn/Ga layered double hydroxides. Industrial & Engineering Chemistry Research. 2018;**57**(48):16264-16271

[89] Li J, Yang W, Wu A, et al. Band-gap tunable 2D hexagonal $(GaN)_{1-x}(ZnO)_x$ solid solution nanosheets for photocatalytic water-splitting. ACS Applied Materials & Interfaces. 2020; **12**(7):8583-8591. DOI: 10.1021/acsami.9b21793

[90] Xu Z, Zhang S, Gao F, et al. Correlations among morphology, composition, and photoelectrochemical water splitting properties of InGaN nanorods grown by molecular beam epitaxy. Nanotechnology. 2018;**29**(47): 475603. DOI: 10.1088/1361-6528/aae0d4

[91] Xu Z, Zhang S, Gao F, et al. Enhanced charge separation and interfacial charge transfer of InGaN nanorods/C_3N_4 heterojunction photoanode. Electrochimica Acta. 2019; **324**:134844. DOI: 10.1016/j.electacta.2019.134844

[92] Xu Z, Zhang S, Liang J, et al. Surface passivation of InGaN nanorods using H_3PO_4 treatment for enhanced photoelectrochemical performance. Journal of Power Sources. 2019;**419**: 65-71. DOI: 10.1016/j.jpowsour.2019.02.050

[93] Lin J, Yu Y, Xu Z, et al. Electronic engineering of transition metal Zn-doped InGaN nanorods arrays for photoelectrochemical water splitting. Journal of Power Sources. 2020;**450**: 227578. DOI: 10.1016/j.jpowsour.2019.227578

[94] Lin J, Yu Y, Zhang Z, et al. A novel approach for achieving high-efficiency photoelectrochemical water oxidation in InGaN nanorods grown on Si system: MXene nanosheets as multifunctional interfacial modifier. Advanced Functional Materials. 2020;**30**:1910479. DOI: 10.1002/adfm.201910479

[95] Wang Y, Wu Y, Sun K, et al. A quadruple-band metal-nitride nanowire artificial photosynthesis system for high efficiency photocatalytic overall solar water splitting. Materials Horizons. 2019;**6**(7):1454-1462. DOI: 10.1039/c9mh00257j

[96] Alotaibi B, Fan S, Nguyen HPT, et al. Photoelectrochemical water splitting and hydrogen generation using InGaN/GaN nanowire arrays. In: 2014 IEEE Photonics Society Summer Topical Meeting Series. IEEE; 2014. pp. 206-207. DOI: 10.1109/sum.2014.112

[97] Wang Y, Vanka S, Gim J, et al. An $In_{0.42}Ga_{0.58}N$ tunnel junction nanowire photocathode monolithically integrated on a nonplanar Si wafer. Nano Energy. 2019;**57**:405-413. DOI: 10.1016/j.nanoen.2018.12.067

[98] Wang Y, Schwartz J, Gim J, et al. Stable unassisted solar water splitting on semiconductor photocathodes protected by multifunctional GaN nanostructures. ACS Energy Letters. 2019;**4**(7): 1541-1548. DOI: 10.1021/acsenergylett.9b00549

[99] Liang J, Han X, Qiu Y, et al. A low-cost and high-efficiency integrated device toward solar-driven water splitting. ACS Nano. 2020;**14**(5):5426-5434. DOI: 10.1021/acsnano.9b09053

[100] Cheng W-H, Richter MH, May MM, et al. Monolithic photoelectrochemical device for direct water splitting with 19% efficiency. ACS Energy Letters. 2018;**3**(8):1795-1800. DOI: 10.1021/acsenergylett.8b00920

Chapter 5

III-Nitride Nanowires: Future Prospective for Photovoltaic Applications

Soumyaranjan Routray and Trupti Lenka

Abstract

Photovoltaic (PV) technology could be a promising candidate for clean and green source of energy. The nanowire technology provides extra mileage over planar solar cells in every step from photon absorption to current generation. Indium Gallium Nitride ($In_xGa_{1-x}N$) is a recently revised material with such a bandgap to absorb nearly whole solar spectrum to increase the conversion efficiency copiously. One of the major technological challenge is in-built polarization charges. This chapter highlights the basic advantageous properties of $In_xGa_{1-x}N$ materials, its growth technology and state-of-the-art application towards PV devices. The most important challenges that remain in realizing a high-efficiency $In_xGa_{1-x}N$ PV device are also discussed. III-Nitride nanowires are also explored in detail to overcome the challenges. Finally, conclusions are drawn about the potential and future aspect of $In_xGa_{1-x}N$ material based nanowires towards terrestrial as well as space photovoltaic applications.

Keywords: III-nitride, polarization charges, efficiency, InGaN/GaN, nanowires, stress, strain

1. Introduction

Photovoltaic (PV) technology is the most emerging way of harnessing huge amount of energy from sun light as compared to solar thermal and photo electrochemical cells [1]. PV devices convert incident photons from sunlight to electricity upon exposed to light. PVs are popular because of its compactness and can be used anywhere for different application [2]. Additionally, involvement of nanostructures further boost the performance of solar cell. Over the past decade, nanostructured solar cell has become hot topics within research community due to its potential to enhance the spectral response of cell. Although, first generation silicon wafer based solar cell leads the current global PV market, however this conventional technology do not have any further scope to improve efficiency and reduce cost [3]. Additionally, it is also not recommended to use Silicon based solar cell for space application due to its low radiation tolerance. Second generation thin film technology such as hydrogenated amorphous silicon (a-Si: H), CIGS, and CdTe could not line-up with wafer-based silicon due to use of rare earth elements and low stability [4, 5]. Furthermore, highly efficient compound semiconductors based third generation solar cell have a demerit of high cost which limits its use in terrestrial applications. Hence, the hunt for low cost high performance solar cell are still unachievable. In the meantime, involvement of nanotechnology could bring a ray of hope for future generation

IntechOpen

solar cell. Nanowire (NW) geometry has remarkable advantages over planner geometry due to optical, electrical, and mechanical effects. New charge separation mechanisms, low defects and low cost also add more mileage to this journey. Looking towards the current scenario, existing PV technologies aren't the solid foundation for the future projection of the renewable energy generation. None of the existing technology can satisfy global energy demand in future [2, 5]. Moreover, if the material or technological limitations restrict the future roadmap of PV technology, then the incorporation of new efficient materials and transpose of technology will be an assurance against high cost and low efficiency solar cells. Newly explored $In_xGa_{1-x}N$ material brings an bunch of opportunities for future PV technology, having capability to absorb full solar spectrum using a single absorber material. One of the major properties of InGaN material is its tunable bandgap from 0.6 to 3.4 eV by changing 'In' content [6–10]. It also has easy growth of nanowire and nanorod structures with proven technology [11–14]. It is a direct bandgap material where photon absorption and direct interband transition can be occurred without interference of phonons to conserve momentum. Additionally, high absorption coefficient of 10^5 is an additional benefit for good absorption with thin layer. Hence the cost can be minimized as well as recombination rate can also be minimized. InGaN also possess a high saturation velocity and a low effective mass of charge carriers, which ensures the more carrier separation along the junction. High radiation tolerance of InGaN are always appreciated for harsh environments. Moreover, InGaN solar cell do not contain any toxic elements such as arsenic, cadmium or phosphorous as used in MJ solar cell. Thus, it is evident that $In_xGa_{1-x}N$ is an extremely allegiant PV material that can enable several photovoltaic devices [15, 16]. It is required to explore state-of-art of different InGaN based PV technologies and new possibilities of InGaN as a hopeful material for future technology [17, 18]. Hence, in this chapter a scope of III-Nitride and its progress with nanostructures have been discussed in order to explore more on future generation solar cell.

2. Planar, nanodisk and nanowire III-nitride solar cells

In the present context mainly different geometry of III-Nitride GaN/InGaN material based solar cell are considered, such as planar, nanodisk and nanowire types. Theoretically, it is anticipated that power conversion efficiency more than 40% could be achievable with GaN/InGaN junctions [19, 20]. However, practically achievable efficiency is quite low [21, 22]. One of the major challenges is the association of in-house defects with InGaN layer at higher 'In' content. Which in turn leads to stuck of immobile charges (known as polarization charges) along interfaces and reduces minority lifetime. Recently III-Nitride nanowires (NWs) are proposed as stand-alone PV devices due to enhanced light trapping, defect and stress-free growth [23, 24]. In general, two widely used structure for nanowire solar cell are (i) axial junction and (ii) radial junction. Axial junction is also known as nanodisk (ND) whereas radial junction is known as core-shell-shell (CSS) solar cell. Vapor–liquid–solid (VLS) technique is one of the popular methods to grow the GaN/InGaN nnaowires [23]. The fabricated NWs have hexagonal cross-section with {0001} orientation of top facet and {1–100}/ {10–10} orientation alongside walls of NWs [23]. Literature shows the comparative analysis of planar and circular NWs reported by Y. Zeng *et al.* [25] and B.M. Kayes *et al.* [26] with achievable conversion efficiency of more than 33% and 50% for Si and GaAs respectively. Additionally, ND and NW SCs are reported by Christesen *et al.* [27], which shows comparative better performance in CSS-NW SCs. The analysis in literature taken circular cross-section into consideration for NWs but in practice,

III-Nitride NWs possess either hexagonal or triangular cross-section. Hence, in this chapter, a scope has been taken to analyze the CSS-NW and ND type SCs with hexagonal crystallographic orientations as per the fabricated devices. The chapter also emphasis on effect of polarization induced electric field with a different 'In' compositions of $In_xGa_{1-x}N$. The schematic of planar, ND and CSS-NW SCs as shown in **Figure 1** and simulated using VICTORY 3D Luminous TCAD. Optical and electrical properties are solved using Poisson's equation, drift diffusion, and photo generation-recombination models. All the structures are designed with $n/i/p$ junctions, which helps to reverse the electric field and match with polarization-induced electric field [28, 29]. A uniform 40 nm and 100 nm thickness are taken for n- and p-GaN layer of three structures. Thin Ni/Au (5 nm/5 nm) layer is used as contact for all three considered cells. Incorporation of Ni layer may reduce series resistance and hence enhance fill factor (FF) of the solar cell. The thickness of intrinsic InGaN layer is varied from 6 nm to 200 nm to optimize the thickness in planar, ND and CSS-NW solar cell. The surface recombination velocity of 10^5 cm/s is taken into consideration for all three structure. The minority carrier lifetime is as follows [29].

$$\tau_{n/p} = \frac{1}{\sigma_{n/p} \times \vartheta_{thermal} \times N_{defect}} + \frac{1}{C_{n/p} \times n} + \frac{1}{C_{Augur} \times n^2} \tag{1}$$

Where $\sigma_{n/p}$ is the capture cross sections of electron or hole, thermal velocity of carriers, $\vartheta_{thermal}$ ($\approx 10^7$cm/s), the defect density (cm^{-2}), N_{defect}, electron capture coefficient for acceptor or donor, $C_{n/p}$, and n is the free electron concentrations.

All the material parameters for GaN and InGaN with different compostion can be found from [30]. The interpolation method is used to calculate all parameters for $In_xGa_{1-x}N$ composite alloys. Defect density in the order of 10^{17}, 10^{14} and 10^{12} are incorporated for planar, ND and CSS-NW SCs respectively [31].

The model developed by Romanov *et al.* [32] and Mastro *et al.* [33] are implemented here to calculate polarization charges along InGaN/GaN heterointerfaces and is expressed as,

$$P_{total} = P_{Lz'}^{pz} + \left(P_L^{sp} - P_T^{sp}\right) cos\varphi \tag{2}$$

Where P_L^{sp} and P_T^{sp} are spontaneous polarizations for top and bottom layer respectively, $P_{Lz'}^{pz}$ is the piezoelectric polarization along the interface which solely depends on strain profile, φ is the angle between planes and basal plane. Angle φ is

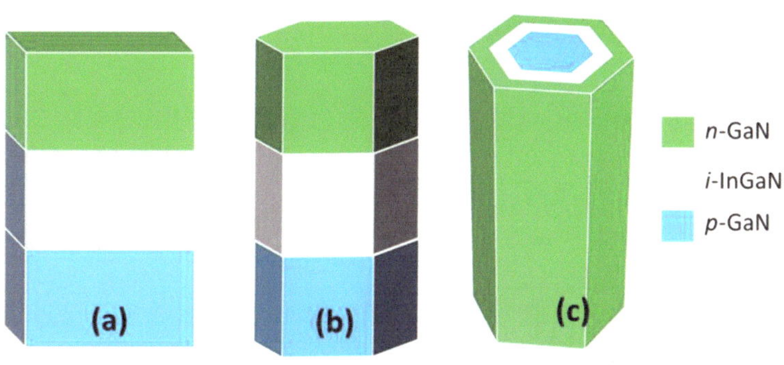

Figure 1.
Schematics of n-i-p *heterojunctions with (a) planar (b) nanodisk (ND) and (c) CSS-NW based solar cell [29].*

considered as 0° for polar {0001} facet and 90° for nonpolar {1–100}/{10–10} facets [33]. Finally, material parameters are incorporated into VICTORY 3D Luminous TCAD to simulate solar cell.

Vertical illumination of light source on the front surface of the device is consider during all analysis as shown in **Figure 1**. Transport of carriers and separation mechanisms in the device depends solely on its geometrical structure. Here, carrier transport in planar, ND and NW type solar cell are explored with the help of energy band diagram. **Figure 2(a)** shows that tilt of energy band along *i*-InGaN region is not in favor of carrier collection due to detrimental polarization effect along {0001} orientation of planar Solar cell. Similarly, ND type solar cell also suffers from polarization effect due to axial growth along {0001} orientation. Hence height of barrier along the junction increases in case of planar and ND solar cell as shown in **Figure 2**.

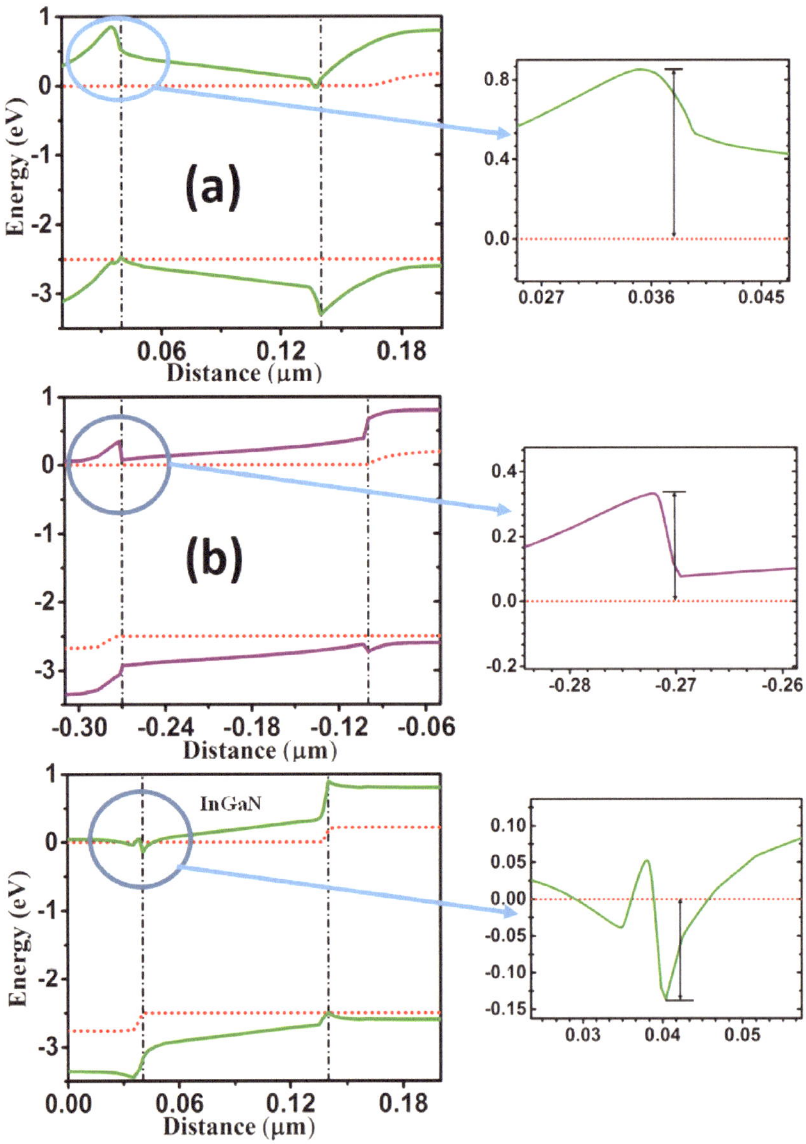

Figure 2.
Energy band diagram of (a) planar (b) ND and (c) CSS-NW solar cell at 100 nm, 150 nm, and 180 nm i-InGaN thickness respectively.

High potential barrier hinders the diffusion of photogenerated carriers to either sides of the contact. Consequently, carrier collection degrades along {0001} interface. Similarly, potential valley is also observed next to barrier along the interfaces, which stocks carriers leading towards low collection.

Thus, with increase in bias voltage, photogenerated carriers accumulate inside the potential valley rather than traveling in i-region. In contrast, CSS-NW type solar cells do not show this effect because of radial separation of carriers. Polarization charges have negligible effect on CSS-NW due to nonpolar facets as shown in **Figure 2(c)**. The inclination energy band in CSS-NW solar cell is in favor of carrier collection. Potential barriers are less as shown in **Figure 2(c)**. Hence a hassle-free movement of carriers is possible in CSS-NW solar cell.

Figure 3 shows dependency of short-circuit current density (J_{sc}) on 'In' composition and thickness of i-InGaN absorber layer. In planar structure, diffusion length carriers are quite low due to high density of defects, which consequently leads to a poor J_{sc}. Thus, higher thickness of i-InGaN do not contribute more towards current due to low diffusion length of carriers. It is also observed that increase in 'In' composition leads to higher polarization charges [24], which also block some of the diffused carriers before collection. Thus J_{sc} shows a poor performance at higher 'In' fraction as shown in **Figure 3**. However, J_{sc} is high in radial junctions as compared to axial junctions. In order to interpret this result, Shockley diode equation is considered for short circuit current as

$$I_{sc} = q \int_{V_{active}} g(\vec{r}) d^3\vec{r} \tag{3}$$

Where volume of the absorption region is, V_{active}, photo-generation rate, $g(\vec{r})$ and position vector is given by $\vec{a}$. The optical photo generation rate, $g(\vec{r})$ is considered as constant g across the entire NW solar cell, which simplifies the complexity of the model.

Thus J_{sc} can be expressed as

$$J_{sc} = q \times g \times L_{active} \tag{4}$$

Where, active absorption layer is, L_{active}. It is noteworthy to mention that due to the structural advantages, CSS-NW possess more than two-fold enhancement in active absorption region, which in turn support more photon absorption and high current. Therefore, it is observed that J_{sc} is quite enhanced in CSS-NW case as compared to ND and planar counterparts. It is due to the high absorption region of CSS structure and higher diffusion length of carriers. Depending on the structural advancement, carrier diffusion length, and lifetime, an optimized 180 nm, 150 nm

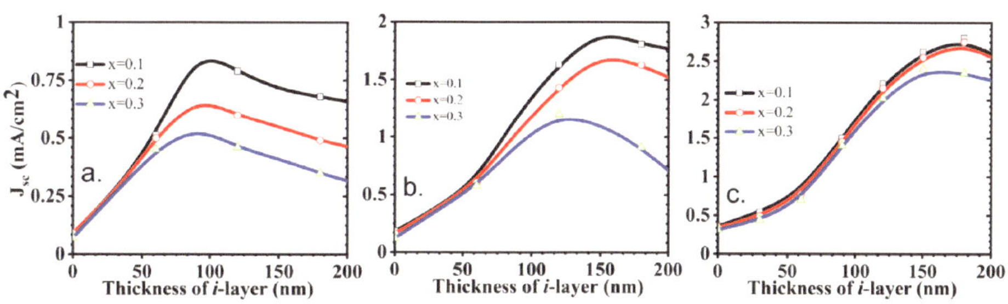

Figure 3.
Short circuit current density (J_{sc}) at 10%, 20%, and 30%, 'In' content for (a) planar (b) Nanodisk and (c) CSS-NW solar cell.

and 100 nm active layers are obtained for CSS-NW, ND and planar structure respectively as given in **Figure 3**.

Figure 4 shows the current density-voltage-power density (J-V-P) curve of n-GaN/ i-$In_{0.1}Ga_{0.9}N$/p-GaN planar, ND and NW type solar cell. The optimized thickness of 100 nm, 150 nm and 180 nm i-InGaN is considered for all performance analysis. It shows a higher J_{sc} in CSS-NW as compared to ND and planar structure. It is anticipated that higher current in CSS-NW solar cell is mainly due to higher active absorption region and efficient carrier separation. It is important to highlight that planar and ND type solar cells shows a stair-case type *J-V* curve, which is not there in CSS-NW. It is may be due to low degree of strain relaxation or higher stress generation along the interface which is again related to structural issues of device. Hence, it is always important to engineer the device structure as per the material properties of the absorber layer. High defect density along the interface is also play a major role for low current in planar and ND solar cell. CSS-NW do not possess a staircase J-V curve due to low defect density, low stress and high degree of strain relaxation. It is also observed that the depth of stair-case in *J-V* curve is increasing with higher 'In' contents. Additionally, higher J_{sc} value of 2.82 mA/cm^2 is noted in CSS-NW solar cell (**Figure 5**).

It is also observed that due to high degree of strain relation, low defects density and more active area of absorption, CSS-NW structure can accommodate higher thickness of active InGaN layer (W) region. Moreover, higher thickness of absorber enhances the probability of more absorption of photons from sunlight. In other hand, active region of planar and ND type solar cells cannot be increased due to are limitation of surface recombination arte, polarization induced electric field, low degree of strain relaxation and defect density. V_{oc} of ND type solar cell is seen to be higher than CSS-NW type solar cell which is may be due to the recombination rate along the junction. However, the rate of increase in J_{sc} of CSS-NW structure is comparatively higher than ND and planar solar cell. Similarly, planar solar cell possesses a low as compared

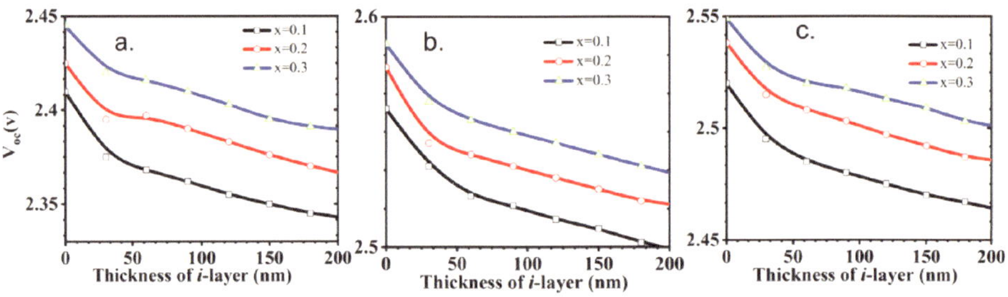

Figure 4.
Open circuit voltage, V_{oc} at 10%, 20%, and 30%, 'In' content for (a) planar (b) Nanodisk and (c) CSS-NW solar cell.

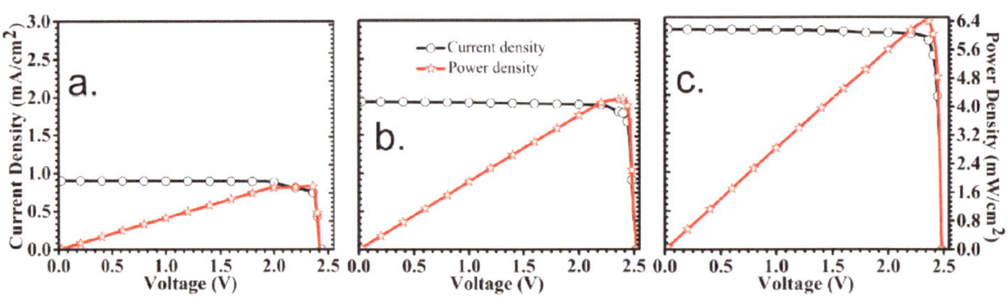

Figure 5.
Current density (black line) – Voltage – Power density (red line) of (a) planar (b) ND and (c) CSS-NW solar cell at 100 nm, 150 nm, and 180 nm i-InGaN thickness respectively.

to ND and NW solar cell. It is mainly due to appearance of staircase type *J-V* curve, which in turn reduces maximum power (P_m) of solar cell. Finally, achievable conversion efficiency (η) of CSS-NW structure is increased by more than 0.5-fold and 2.5-fold, compared to ND and planar type solar cells respectively. More analysis on different types of CSS nanowire is also studied and can found in [34–36].

	W (nm)	J_{sc} (mA/cm^2)	V_{oc} (V)	FF (%)	Efficiency η (%)
Planar	100	0.90	2.41	82.7	1.8
Nanodisk (ND)	150	1.93	2.52	87.6	4.28
CSS-Nanowire (NW)	180	2.84	2.47	91.6	6.46

3. Conclusion

In this chapter, the importance of nanowire solar cell with III-Nitride material is explored in a detailed manner. A comparative analysis is carried out with planar, nanodisk and nanowire type solar cell and concluded that nanowire type structure shows a better performance as compared to others. Additionally, it is found that nanowires in InGaN materials are grown either in triangular or hexagonal orientation. The strain relaxation is more in CSS-nanowires which in turns leads to low in-house defect density along the interfaces. CSS-NWs are also able to accommodate higher thickness of intrinsic material due to high carrier diffusion length. Radial separation of carriers also provides more surface area and better control on carrier separation mechanisms. Hence, it is concluded that radial growth of nanowire provides a broad range of opportunity for performance enhancement of solar cell. The similar type of observation are also applicable to LASER and light emitting diodes, where III-Nitride materials are used.

Acknowledgements

The authors acknowledge Dept. of Science and Technology, Govt. of India for financial support for providing the instrumental facility under DST-FIST in Dept. of ECE, SRM Institute of Science and Technology, Chennai, India.

The authors also acknowledge Visvesvaraya PhD Fellowship of MeitY, Govt. of India and TEQIP-II for procurement of Silvaco TCAD tool in Department of ECE, NIT Silchar for carrying out the research work.

Conflict of interest

The authors declare that they have no known competing financial interests or personal relationships that could have appeared to influence the work reported in this chapter.

Appendices and nomenclature

InGaN	indium gallium nitride
GaN	gallium nitride
CIGS	copper indium gallium selenide
CdTe	cadmium telluride

Author details

Soumyaranjan Routray[1*] and Trupti Lenka[2]

1 SRM Institute of Science and Technology, Tamil Nadu, India

2 National Institute of Technology Silchar, Assam, India

*Address all correspondence to: s.r.routray@ieee.org

References

[1] C. S. olanki, "Solar Photovoltaics Fundamentals, Technologies and Application", PHI Learning Pvt. Ltd, New Delhi, 2012.

[2] IPCC, 2019: Summary for Policymakers. In: Climate Change and Land: an IPCC special report on climate change, desertification, land degradation, sustainable land management, food security, and greenhouse gas fluxes in terrestrial ecosystems [P.R. Shukla, J. Skea, E. Calvo Buendia, V. Masson-Delmotte, H.- O. Pörtner, D. C. Roberts, P. Zhai, R. Slade, S. Connors, R. van Diemen, M. Ferrat, E. Haughey, S. Luz, S. Neogi, M. Pathak, J. Petzold, J. Portugal Pereira, P. Vyas, E. Huntley, K. Kissick, M. Belkacemi, J. Malley, (eds.)]. In press.

[3] M. T. Kibri, A. Ahammed, S. M. Sony, F. Hossain, S. U. Islam "A Review: Comparative studies on different generation solar cells technology. Proceedings of 5th International Conference on Environmental Aspects of Bangladesh, 11-12 Sep, 2015.

[4] Y. Hamakawa, "Thin-Film Solar Cells: Next Generation Photovoltaics and Its Applications," Spinger, New York, 2004.

[5] M. A. Green, K. Emery, Y. Hishikawa, W. Warta, E. D. Dunlop, D. H. Levi, A. W. Y. Ho-Baillie, Solar cell efficiency tables (version 49). Prog. Photovolt: Res. Appl., vol. 25, no. 1, pp. 3-13, 2017.

[6] V. Y. Davydov et al., "Absorption and Emission of Hexagonal InN. Evidence of Narrow Fundamental Bandgap," Physical Status Solidi B, vol. 229, no. 3, pp.R1-R3, 2002.

[7] J. Wu, W. Walukiewicz, K.M. Yu, J.W. Ager, E.E. Haller, and Schaff, W.J. Hai Lu, "Unusual properties of the fundamental band gap of InN," Applied Physics Letters, vol. 80, no. 254, pp.4741-3, 2002.

[8] T. Matsuoka, H. Okamoto, M. Nakao, H. Harima, and E. Kurimoto, "Optical bandgap of wurtzite InN," Applied Physics Letters, vol. 81, no. 7, pp.1246-8, 2002.

[9] I. Vurgaftman, and J.R. Meyer, "Band parameters for nitrogen-containing semiconductor," Journal of Applied Physics, vol. 94, no. 6, pp. 3675-96, 2003.

[10] W. Wu, W. Walukiewicz, "Band gaps of InN and group III nitride alloys," Superlattices and Microstructures, vol. 34, pp. 63-75, 2003.

[11] Y. Li, J. Xiang, F. Qian, S. Gradečak, Y. Wu, H. Yan, D. A. Blom, and C. M. Lieber, "Dopantfree GaN/AlN/AlGaN radial nanowire heterostructures as high electron mobility transistors," Nano Letters, vol. 6, no. 7, pp. 1468-1473, Apr. 2006.

[12] H. Zhang, A. Messanvi, C. Durand, J. Eymery, P. Lavenus, A. Babichev, F. H. Julien, and M. Tchernycheva, "InGaN/GaN core/shell nanowires for visible to ultraviolet range photo detection," Phys. Status Solidi Appl. Mater. Sci., vol. 940, no. 4, pp. 936-940, Jan. 2016.

[13] M. Tchernycheva, P. Lavenus, H. Zhang, A. V. Babichev, G. Jacopin, M. Shahmohammadi, F. H. Julien, R. Ciechonski, G. Vescovi, and O. Kryliouk "InGaN/GaN Core–Shell Single Nanowire Light Emitting Diodes with Graphene-Based P-Contact," Nanoletters, vol. 14, pp. 2456-2465, Apr. 2014.

[14] F. Qian, Y. Li, S. Gradecak, H. G. Park, Y. Dong, Y. Ding, Z. L. Wang, and C. M. Lieber, "Multiquantum-well nanowire heterostructures for wavelength-controlled lasers," Nat.

Mater., vol. 7, no. 9, pp. 701-706, Aug. 2008.

[15] D. V. P. Mclaughlin and J. M. Pearce, "Progress in Indium Gallium Nitride Materials for Solar Photovoltaic Energy Conversion," Met. Mat Trans A, vol. 44, no. 44, pp. 1947-1954, Feb.2013.

[16] J. Wu, W. Walukiewicz, K. M. Yu, W. Shan, J. W. Ager, E. E. Haller, H. Lu, W. J. Schaff, W. K. Metzger, and S. Kurtz, "Superior radiation resistance of In 1-xGa xN alloys: Full-solar-spectrum photovoltaic material system," J. Appl. Phys., vol. 94, no. 10, pp. 6477-6482, Nov. 2003

[17] A. G. Bhuiyan, K. Sugita, A. Hashimoto, and A. Yamamoto, "InGaN solar cells: Present state of the art and important challenges," IEEE J. Photovoltaics, vol. 2, no. 3, pp. 273-296, July 2012.

[18] J. Wu, "When group-III nitrides go infrared: New properties and perspectives," J. Appl. Phys., vol. 106, no. 1, pp.011101-1-011101-28, 2009.

[19] H. Hamzaoui, A. S. Bouazzi, and B Rezig, "Theoretical possibilities of $In_xGa_{1-x}N$ tandem PV structures," Sol. Energy Mater. Sol. Cells, vol. 87, no. 1-4, pp. 595-603, 2005.

[20] J. Y. Chang, S. H. Yen, Y. A. Chang, B. T. Liou, and Y. K. Kuo, "Numerical investigation of high-efficiency InGaN-based multijunction solar cell," IEEE Trans. Electron Devices, vol. 60, no. 12, pp. 4140-4145, 2013.

[21] J. P. Shim, S. R. Jeon, Y. K. Jeong, and D. S. Lee, "Improved efficiency by using transparent contact layers in InGaN-based p-i-n solar cells," IEEE Electron Device Lett., vol. 31, no. 10, pp. 1140-1142, 2010.

[22] R. M. Farrell, C. J. Neufeld, S. C. Cruz, J. R. Lang, M. Iza, S. Keller, S. Nakamura, S. P. Denbaars, U. K. Mishra, and J. S. Speck, "High quantum efficiency InGaN/GaN multiple quantum well solar cells with spectral response extending out to 520 nm," Appl. Phys. Lett., vol. 98, no. 20, pp. 2009-2012, 2011.

[23] A. Messanvi, H. Zhang, V. Neplokh, F. H. Julien, F. Bayle, M. Foldyna, C. Bougerol, E. Gautier, A. Babichev, C. Durand, J. Eymery, and M. Tchernycheva, "Investigation of Photovoltaic Properties of Single Core-Shell GaN/InGaN Wires," ACS Appl. Mater. Interfaces, vol. 7, no. 39, pp. 21898-21906, 2015.

[24] T. M. G. Sarwar and R. C. Myers, "Exploiting piezoelectric charge for high performance graded InGaN nanowire solar cells," Appl. Phys. Lett., vol.101, no. 14, pp. 143905-1-5, 2012.

[25] Y. Zeng, Q. Ye and W. Shen, "Design principle for single standing nanowire solar cells: going beyond the planar efficiency limits," Scientific reports, vol. 4, no.11, pp. 1-7, 2014.

[26] M. Kayes and H. A. Atwater, "Comparison of the device physics principles of planar and radial p-n junction nanorod solar cells," J. Appl. Phys., vol. 97, no. 11, pp. 114302-1-114302-11, 2005.

[27] J. D. Christesen et al., "Design principle for photovoltaic devices based on Si nanowires with axial or radial p-n junctions," Nanoletters, vol. 12, no. 11, pp. 6024-6029, 2012.

[28] J. Y. Chang and Y. K. Kuo, "Numerical Study on the Influence of Piezoelectric Polarization on the Performance of p-on-n (0001)-Face GaN/InGaN p-i-n Solar Cells," IEEE Electron Device Lett. , vol. 32, no. 7, pp. 937-939, 2011.

[29] S. R. Routray, T. R. Lenka, "Performance Analysis of Nanodisk and Core/Shell/Shell-Nanowire Type

III-Nitride Heterojunction Solar Cell for Efficient Energy Harvesting," Superlattices and Microstructures (Elsevier), vol.111, no. 12, pp.776-782, Nov. 2017

[30] J. Wu, "When group-III nitrides go infrared: New properties and perspectives," J. Appl. Phys, vol. 106, no. 1, pp. 011101-1-011101-28, 2009.

[31] K. L. Kavanagh, "Misfit dislocations in nanowire heterostructures," Semicond. Sci. Technol., vol. 25, no. 2, pp. 1-7, 2010.

[32] A. E. Romanov, T. J. Baker, S. Nakamura, and J. S. Speck, "Strain-induced polarization in wurtzite III-nitride nonpolar layers," J. Appl. Phys., vol. 100, no. 2, pp. 23522-1-10, 2006.

[33] M. A Mastro, B. Simpkins, G. T. Wang, J. Hite, C. R. Eddy, H.-Y. Kim, J. Ahn, and J. Kim, "Polarization fields in III-nitride nanowire devices," Nanotechnology, vol. 21, no. 14, pp. 145205, 2006

[34] S. R. Routray, B. Shougaijam, T. R. Lenka, "Exploiting Polarization Charge for High Performance (000-1) facet GaN/$In_xGa_{1-x}N$ Based Triangular Nanowire Solar Cell," *IEEE Journal of Quantum Electronics*, vol. 53, no. 5, pp. 1-8, July 2017.

[35] S. R. Routray, T. R. Lenka, "InGaN-based Solar Cell: A Wide Solar Spectrum Harvesting Technology for 21st Century" *CSI Transactions on ICT* (Springer), vol. 6, no. 1, pp. 83-96, October 2017

[36] S. R. Routray, T. R. Lenka, "Polarization Charges in High Performance GaN/InGaN Core/Shell Multiple Quantum Well Nanowire for Solar Energy Harvesting" *IEEE Trans. on Nanotechnology*, vol. 17, no. 6, pp. 1118-1124, June 2018.

Chapter 6

In Situ TEM Studies of III-V Nanowire Growth Mechanism

Carina B. Maliakkal

Abstract

Growing nanowires inside a transmission electron microscope (TEM) and observing the process in situ has contributed immensely to understanding nanowire growth mechanisms. Majority of such studies were on elemental semiconductors – either Si or Ge – both of which are indirect bandgap semiconductors. Several compound semiconductors on the other hand have a direct bandgap making them more efficient in several applications involving light absorption or emission. During compound nanowire growth using a metal catalyst, the difference in miscibility of the nanowire species inside the metal catalyst are different, making its growth dynamics different from elemental nanowires. Thus, studies specifically focusing on compound nanowires are necessary for understanding its growth dynamics. This chapter reviews the recent progresses in the understanding of compound semiconductor nanowire growth obtained using in situ TEM. The concentrations of the nanowire species in the catalyst was studied in situ. This concentration difference has been shown to enable independent control of layer nucleation and layer growth in nanowires. In situ TEM has also enabled better understanding of the formation of metastable crystal structures in nanowires.

Keywords: compound nanowire, transmission electron microscopy, ledge-flow, semiconductor, GaAs, in situ techniques, wurtzite, zincblende, polytypism

1. Introduction

The high surface-to-volume ratio and the high aspect ratio of the nanowire geometry paves the way to a plethora of interesting advantages. Growing materials as nanowires has enabled the formation of metastable crystal phases, in turn enabling crystal structure tuning [1–4]. Integration of different lattice-mismatched materials into the same structure was also achieved; compared to growth of hetero-epitaxial films, defect-free growth is easier in nanowires because of the small diameter (a few 10 or 100 nanometers) and small interfacial area [5–7]. Yet another advantage of nanowire growth is to form alloy compositions which are unstable in the bulk phase [8]. Materials generally grown in the nanowire morphology can be broadly classified as elemental and compound [9]. Metallic nanowires (Ti, Fe, Co, Ni, Sn etc.) and elemental semiconductors (e.g. Si, Ge) fall under the category of elemental materials. Stochiometric compound nanowires are either compound semiconductors (e.g. GaAs, ZnO) or insulators (e.g. Al_2O_3, Si_3N_4). Alloy nanowires are also possible, e.g. Si_xGe_{1-x}, $Al_xGa_{1-x}As$. Controlling the electronic, bandgap-engineering related, structural, compositional, morphological, mechanical and optical properties of semiconductor nanowires enables its application in devices such solar cells, [10–13] electronics, [14–16] LEDs, [16, 17] LASERS, [18, 19] photodetectors, [20, 21] thermoelectrics, [22, 23] biosensors [24, 25] and qubits [26–29].

Nanowires can be fabricated by a top-down approach (where regions of a film are selectively etched) or by a bottom-up approach (where the nanowires are grown on a substrate) [30]. Top-down approach often result in rough defected surfaces [9]. One way to grow nanowires bottom-up is by electrochemical deposition [31–34]. The bottom-up growth of nanowires from a gas phase precursor supply is what we will discuss in more detail here. The bottom-up nanowires growth from gas phase are done either with or without a foreign metal catalyst. The nanowire growth using a metal catalyst was proposed to proceed by the vapor–liquid–solid or 'VLS' mechanism [35]. According to the VLS mechanism, the nanowire elements or their precursor supplied in the vapor (V) phase gets dissolved in the liquid (L) 'catalyst' and after supersaturation precipitates out as the solid (S) nanowire (**Figure 1**). The metallic liquid, in addition to providing a nucleation point for the solid nanowire, fosters the gathering and in some cases the decomposition of precursors – hence often called 'catalyst' [9]. Later a similar growth mode called the vapor-solid–solid (VSS) was also proposed for when the catalyst is a solid, instead of the liquid catalyst in VLS [36, 37]. When there is no foreign catalyst used the nanowire growth can proceed in either of the two ways: (i) a self-catalyzed mode where the metallic element of the nanowire forms the liquid catalyst droplet [38–40] or (ii) a non-catalyzed vapor-solid route where the material from the vapor phase directly attaches to the solid nanowire without any liquid or solid catalyst [41]. The yield of nanowire growth without an external catalyst can be increased by the use of selective-area dielectric mask to keep some areas unfavorable for nucleation; small openings in the mask acting as preferential nucleation site for nanowires [42, 43].

Some of the common techniques for growing nanowires with gas phase precursors by the aforementioned mechanisms include chemical vapor deposition (CVD), metalorganic CVD (MOCVD) and molecular beam epitaxy (MBE). These systems were initially designed for growing thin films and later adapted for growing nanowires. Usually for growing nanowires in any of these systems the catalyst-coated substrate is loaded into the system, the system is closed and precursors are supplied at

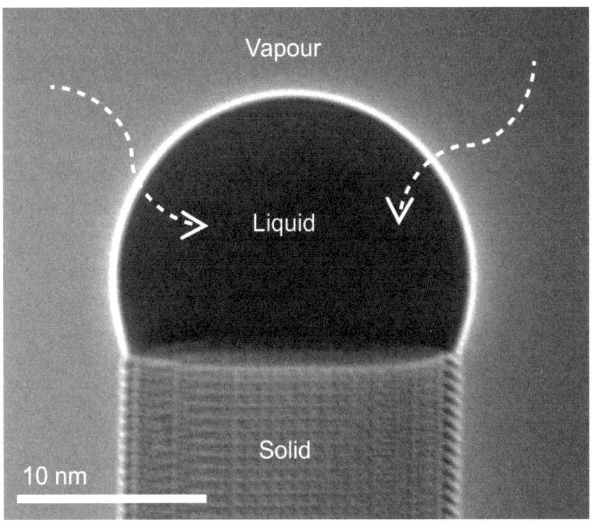

Figure 1.
Nanowire growth with a liquid catalyst is explained by the VLS mechanism. Accordingly, the supplied vapor phase precursor species dissolves in the liquid catalyst and at appropriately high supersaturation crystallizes atomic layers of the solid nanowire. The TEM image shown here was captured in situ while an atomic layer was growing.

appropriate temperature and pressure. In conventional growth systems, either there is no in situ monitoring during growth or there is some large-area indirect monitoring. MBE systems sometimes monitors the crystal structure of the surface layer by RHEED (reflection high-energy electron diffraction). Some MOCVD systems are equipped with in situ light reflectance monitors which can be used to estimate the increase in sample height and surface roughening. These methods are used conventionally for tracking growth of thin films from large areas of the sample. Using these techniques for monitoring nanowires demand some modifications. After growing nanowires, the samples can be elaborately analyzed ex situ by methods relevant to the study. Typical characterization techniques are scanning electron microscopy (SEM), transmission electron microscopy (TEM), X-ray diffraction (XRD), energy dispersive X-ray spectroscopy (XEDS), photoluminescence, etc.

Such ex situ characterizations could suffice for studying nanowire morphology, crystallinity and composition. Analysis of nanowires from multiple growths with different parameters can help indirectly understand the growth mechanism to some extent. Still the dynamics of the growth understood based on such ex situ characterizations are largely speculative. Moreover, some attributes of the nanowire could be different while growing (at high temperature with precursor supply) and after cooling down the system for post mortem analysis. Monitoring nanowire growth in situ certainly has advantages in elucidating the growth mechanism and dynamics. An example of a phenomenon which was discovered only due to in situ observation of individual nanowires is truncation — where the nanowire-catalyst interface has a dynamic non-flat surface near the triple-phase-line [44–48]. (Triple-phase-line refers to the periphery of the interface between the nanowire and the catalyst droplet where the vapor, liquid and solid phases meet.) Another interesting in situ observation was that the nucleation of wurtzite layer happens at, or at least very close to, a corner of the triple-phase-line (observation of the precise location being elusive due to the 'limited' temporal resolution compared to the expected extremely rapid growth of the nucleus to beyond the critical size) [49].

2. In situ techniques

Observing and characterizing the nanowires while they are growing is called in situ growth monitoring. Strictly speaking, 'in operando' is the exact word, but we stick to 'in situ' to conform to popular usage. In situ techniques can provide directly interpretable and time-resolved observations enabling better understanding of the growth mechanism, which in turn empowers better control of nanowire growth for specific technological applications.

In situ characterization of nanowire crystal structure or nanowire morphology has been reported using various techniques. In situ RHEED attached to MBE systems can be used to follow crystal structure changes and nucleation/birth of ensemble of nanowires [50, 51]. By modifying the optical reflectometry techniques that have been used conventionally in MOCVD systems, the nanowire diameter and length evolution has been monitored in situ in real time for an ordered array of nanowires [52]. Combining finite difference frequency domain simulations with in situ reflectometry enabled monitoring growth of randomly positioned nanowires (i.e. periodic array was not a necessity) [53]. In situ X-ray diffraction (XRD) has been used to study crystal phase of the nanowire [54, 55] and the catalyst phase [56]. In situ infrared spectroscopy has been used to correlate surface chemistry during nanowire growth to its morphology [57–59] or the choice of growth direction [60]. Line-of-sight quadrupole mass spectrometry in situ was used to study different stages of nanowire growth including nanowire nucleation [61]. All these techniques give ensemble averaged results.

In situ imaging techniques on the other hand allows monitoring individual nanowires. Optical microscopes due to the limited spatial resolution are not ideal for observing growth evolution of nanowire (though some studies have been attempted using confocal optical microscopy using photoluminescence measurements [62]). Scanning electron microscopes (SEMs) have better spatial resolution than optical microscopes and could be used to monitor nanowire growth [63–67]. In situ SEM combined with Auger electron spectroscopy has been used to correlate nanowire growth and morphology to surface chemistry [63]. In situ electron backscattered electron diffraction (EBSD) performed during growth in an SEM has been used to study crystal phases and crystallographic orientation [64]. An SEM uses electron scattering from a sample while a transmission electron microscope (TEM) uses the electrons transmitted through a thin sample (preferably less than ~50 nm) to form images. TEMs have better spatial resolution than SEMs. Be it in an in situ SEM or TEM study, a video or a series of images are captured to study the dynamics of the process in relation with the specimen environment. A key advantage of using in situ microscopic techniques, particularly in situ TEM, is that localized or dynamic behavior happening at individual wires could be investigated. One limitation to studying nanowire growth inside a microscope is that electron microscopes require vacuum environment to minimize electron scattering in the air outside the specimen. So, often the growth conditions, e.g. pressure, used for the in situ growth study are slightly modified compared to a conventional growth method. Typical total pressures used in conventional ex situ CVD are much beyond the maximum attainable pressure for in situ TEM experiments. Majority of the pressure in the ex situ CVD case is from the carrier gas. By careful design of the TEM and the growth chamber, it is in principle possible to obtain comparable precursor partial pressures.

3. Techniques for growth nanowire by in situ TEM

The very first demonstration of nanowire growth by in situ TEM was from Prof. Yang's group which validated the VLS mechanism experimentally using the Au-catalyzed Ge nanowire growth [68]. This seminal experiment was conducted by heating Au nanoclusters along with micrometer-sized Ge particles. They neither used a continuous supply of Ge-precursor nor a closed system. Over the course of time technological advances in the field of TEM paved the way to environmental TEM (ETEM), where the pressure near the sample can be orders of magnitude higher than a conventional TEM. Studies in which a continuous supply of precursors was used were reported [41, 46–49, 69–90].

However, an ETEM is not a necessity for studying CVD nanowire growth in situ by TEM, it is possible by using a closed or isolated cell instead. In principle, it is possible that a cell isolated from the microscope vacuum is used; gaseous precursors can be supplied continuously to this cell by external inlet gas-tubes and removed by outlet tubes, without releasing the gases to the microscope environment. Another strategy is to use completely closed cells, in which powders of precursor material are deposited in the cell and then sealed [91]. These powders are heated intentionally to evaporate it so as to form a vapor-phase supply of precursors to the catalyst for growth [91]. An intermediate method, which is feasible with commercially available instruments for gas handling, is to pre-deposit powdered material on the isolated cell but externally supply carrier gases such as H_2 or N_2 (no gases are released here to the microscope environment) [92]. The cells have an electron transparent amorphous film both at the top and at the bottom of the cell. An advantage of this strategy is that any ordinary TEM can be used for it. However, the thickness of the top and bottom casing combined could be substantial, reducing the attainable spatial resolution.

In the more conventional open heating cell geometry, there is either one layer of amorphous layer or none, providing better spatial resolution. Commercial chips are available with a few holes made in a thin amorphous film. When a piece of commercially available substrate wafer is loaded vertically in the TEM [69, 70] or a lithographically patterned cantilever chips is used [76, 93] for growing epitaxial aligned nanowires there is no film on top or bottom of the nanowire sample, enabling epitaxial growth and better spatial resolution; in such cases the resolution of the microscope and thickness of the sample would be the bottleneck. TEM resolution is currently restricted by technical limitations, not by the physically attainable limit; over the years TEM resolution has been constantly improving and this evolution is visible if we look at reports of nanowire growth with in situ TEM as well.

4. In situ TEM of elemental semiconductor nanowire growth

Si and Ge nanowire growth has been extensively studied by in situ TEM [46, 68–78, 81, 82, 85–88]. Several aspects such as diameter dependance of growth kinetics [70], nucleation kinetics [87], surface faceting [69], surface migration of catalyst (Au) on nanowire (Si) surface [71], tapering [94], and kinking [75] have been investigated. Depending on the growth conditions such as temperature, catalyst particle and precursor pressures the growth proceeds either by the VLS mode [46, 68–72, 74–76, 78, 81, 82, 85–88] or the VSS mode [46, 72–74, 77, 78, 82, 88]. It is interesting to note that VLS growth has been observed to occur even below the eutectic temperature [72].

The nanowire catalyst interface is atomically flat, except when a ledge is growing. The layer-by-layer growth of nanowire atomic layers has been studied in situ during the VLS growth of elemental nanowires [74, 78]. A new (bi)layer starts only after the previous one is completely grown (at least for the nanowire diameters studied) [74, 78]. The time each layer takes to complete once it has nucleated can be called ledge-flow time (or layer completion time, also called step-flow time in some references). We will use the term incubation time for the difference between the ending of one layer and the start of the next layer. (This is not to be confused with the incubation time before the birth/nucleation of the nanowire itself). In VLS growth of elemental nanowires each layer grows instantaneously (ledge-flow time ~ 0) while there is a significant incubation/waiting time between successive layer-growth events [74, 78]. This observation can be explained by a very simple argument — the amount of material required to raise the chemical potential high enough to nucleate a layer is sufficient for forming one full layer as soon as it nucleates. So the layer grows rapidly once nucleated [74]. There is a considerable incubation time, which in turn determines the average nanowire growth rate.

Most theoretical models for nanowire growth kinetics assume instantaneous layer completion and the growth rate is calculated in a nucleation-limited regime [95–98]. This assumption seems to be valid for the VLS growth of elemental nanowires we discussed above. However, we will now discuss in this section about elemental nanowires and the next section about compound nanowires cases where this assumption of instantaneous layer-growth breaks down.

As mentioned before, the growth can proceed by the VSS route where the catalyst is a solid particle. In the VSS growth of elemental nanowires the layer completion is slow [73, 74, 78, 82]. The incubation time in the VSS case is shorter than in VLS [74, 78]. The solubility of the growth material in the solid catalyst is much lower than in a liquid catalyst, thus a small amount of excess species can increase the chemical potential sufficiently to nucleate a new layer — making the incubation time short [74, 78]. But the limited amount of material present could be insufficient for forming a complete bilayer, in turn making the ledge-flow process slow [74, 78]. The limited solubility of the nanowire species inside the solid catalyst offers the opportunity to grow compositionally abrupt axial heterostructure [74]. Another interesting aspect about VSS growth of elemental nanowires is that there can be two or more ledges growing simultaneously [73, 78, 82]. This also is in contrast to VLS growth of elemental nanowires were a second ledge starts only after the first is fully grown.

5. In situ compound nanowire growth

Compound nanowires grown inside TEM include insulator materials (Al_2O_3 [45]) and semiconductors (GaAs [49, 84, 99], GaN [79, 83, 100], GaP [46, 80], InAs [91] and PdSe [92]). VSS growth of compound nanowires in a TEM with a supply of precursors has not been reported so far; hence the discussion we have in this section is restricted to VLS growth of compound nanowires. In the cases where atomic resolution videos where obtained, ledge-flow was not instantaneous [49, 83, 84, 90]. The initial studies of MOCVD combined with in situ TEM were at very low precursor pressures compared to the typical 'ex situ' MOCVD; [83, 84] hence it was not sure if the gradual ledge-flow was representative of ex situ growths as well. The latest report was with orders of magnitude higher pressures than previous studies, but still the precursor pressures values were on the lower end of conventional ex situ MOCVD growth parameter regime [90]. If or not the ledge-flow of atomic layers is gradual in the entire range of growth parameters used in ex situ growths is yet to be investigated.

The gradual ledge-flow growth in compound nanowires, is in striking contrast to the VLS monoatomic nanowire growth. But this difference between elemental and compound nanowires is simple to understand. In elemental nanowire only one material species controls both nucleation and layer-growth events. For example, during Si nanowire growth with a Au catalyst the Si dissolving in the Au is the key factor. At typical growth temperatures of Si nanowire growth (400–600°C) the liquidus line where the Au-Si system is at equilibrium is with about 20–28% Si (depending on the growth temperature). A little extra Si is insufficient to supersaturate the system enough to trigger a nucleation event. The amount of excess Si that accumulates during the incubation time and triggers the nucleation of a layer could thus suffice to form an entire layer. However, in a compound nanowire case the miscibility of two different nanowire species within the catalyst could be decisive, in turn making the dynamics more complex. Species like Ga, In, Al and Zn alloys readily with Au while species like As, N, P and O are hardly soluble in Au [101]. In the case of Au-catalyzed GaAs growth, for example, theoretical calculations predicted that Ga mixes readily in Au but As has poor solubility in Au [102, 103]. Experimental studies of the catalyst composition was mostly done ex situ post growth until very recently.

6. Catalyst composition measured in situ

Understanding the concentration of the different species during growth is key to understanding the growth dynamics. Parameters like surface and interface energies, vapor pressure and chemical potential of the catalyst are dependent on the catalyst composition [104–106]. The catalyst composition measured ex situ depends not just on the growth parameters, but also on the conditions used to terminate growth and cool down the sample (the ambient gas, ramp down rate etc.) [2, 44, 107]. Typically in MOCVD the temperature is decreased from the growth temperature down to either room temperature or an intermediate temperature in a group V (or group VI) precursor environment to prevent etching and surface roughening. At the initial part of this ramp down, where the temperature is still adequate to grow, the group III (or II) species already present in the catalyst reacts with the group V precursor to form an additional nanowire segment, in turn decreasing the concentration of the group III species in the catalyst [2, 44, 107]. Hence in situ measurement is key.

Recently in situ measurement of the catalyst composition during the growth process was reported [89]. X-ray energy dispersive spectroscopy (XEDS) spectroscopy was used in situ to study the catalyst composition during Au-catalyzed GaAs growth performed inside an ETEM. Trimethylgallium (TMGa) and arsine (AsH_3) were used as the precursors. The XEDS measurement was conducted in the TEM mode by condensing the beam to a small region and positioning it in the front part of the catalyst (like depicted in **Figure 2a**). Since the nanowire was growing, the sample stage was constantly repositioned so that the beam is all the time on the catalyst itself, and not hitting the nanowire part. The XEDS signal from Au, Ga and As was studied. The catalyst had a significant amount of Ga alloyed with the Au. The Ga % in the catalyst was found to increase with both temperature and the Ga precursor flux. **Figure 2b** shows the Ga % as function of the V/III ratio i.e. the ratio of the group V precursor to the group III precursor. These experiments were done in the 420–500°C temperature range. At these temperatures, the catalyst interaction with nanowire depends on the TMGa flow – (a) in the absence of a TMGa flow the catalyst particle can etch the nanowire (similar to what was reported by Tornberg *et al.* [108]); (b) at an intermediate TMGa flow there is neither growth nor etching; (c) at a slightly higher TMGa there is nanowire growth where the Ga % increases with increasing TMGa flow to a quasi-steady state and the catalyst bulges due to the additional Ga; (d) eventually there is a regime with truncated nanowire-catalyst interface and (e) finally at even higher TMGa the catalyst bulges and topples.

The As signal in the EDX spectra was too low to be conclusively attributed to be arising from the catalyst and was suspected to be due to scattered signal from the nanowire [89]. The As content was however estimated by an indirect method — calculating phase diagrams or liquidus lines for different As % and comparing the Ga % in these calculations to the measured Ga % value. The estimated minimum As % in the catalyst was ~0.01%. For the nanowire dimensions used, this would be

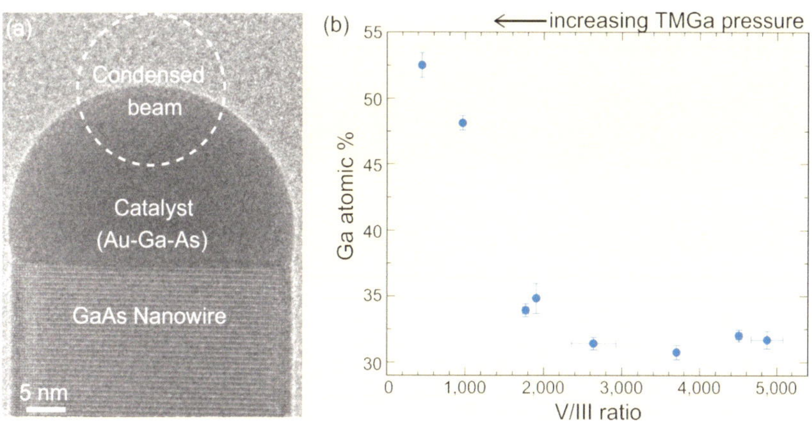

Figure 2.
In situ catalyst composition measurement. (a) TEM image of a nanowire. The catalyst composition was measured in situ by XEDS by condensing the beam in the front of the catalyst. (b) the Ga % in the catalyst is plotted as a function of the V/III ratio. The As % measured in the catalyst was negligible. (a) and (b) are adapted from Maliakkal et al. *2019 [89] with permission as per creative commons license.*

less than the amount of As required to form one complete bilayer. Indirect estimates based on ex situ growth, [95] phase diagrams, [109] and theoretical calculations related to Au-catalyzed GaAs nanowire growth [102, 103] also suggested low As solubility in the Au-Ga alloy.

7. Independent control of layer nucleation and layer-growth

The concentration difference between the two different nanowire species in the catalyst [89] implies that both these species could affect the growth in different ways. This was studied in detail, again using Au-catalyzed GaAs (and TMGa and AsH_3 as precursors) [90]. Two sets of experiments were investigated – one was a TMGa series (where AsH_3 and temperature were kept constant) and a second was AsH_3 series (where TMGa and temperature were kept constant). The ledge-flow time was found to decrease drastically with increasing AsH_3 flow (**Figure 3b**); thus the ledge-flow process was understood to be limited by the As availability. This agrees well with the low As % present in the catalyst [89]. (The idea that ledge-flow is limited by As availability was proposed in an earlier study [84] but not elaborately investigated there.) On increasing TMGa flow in a separate experiment the incubation time decreased drastically while the ledge-flow time remained rather unchanged (**Figure 3c**). This indicated that nucleation of a new layer is triggered by excess Ga.

The experimental observations for the TMGa and AsH_3 series matched stochastic Monte Carlo simulations done based on mass transport and nucleation theory [89]. An example of how As % and Ga % in the catalyst varies in an almost cyclic way during the simulated layer-growth cycle is shown in **Figure 3d**.

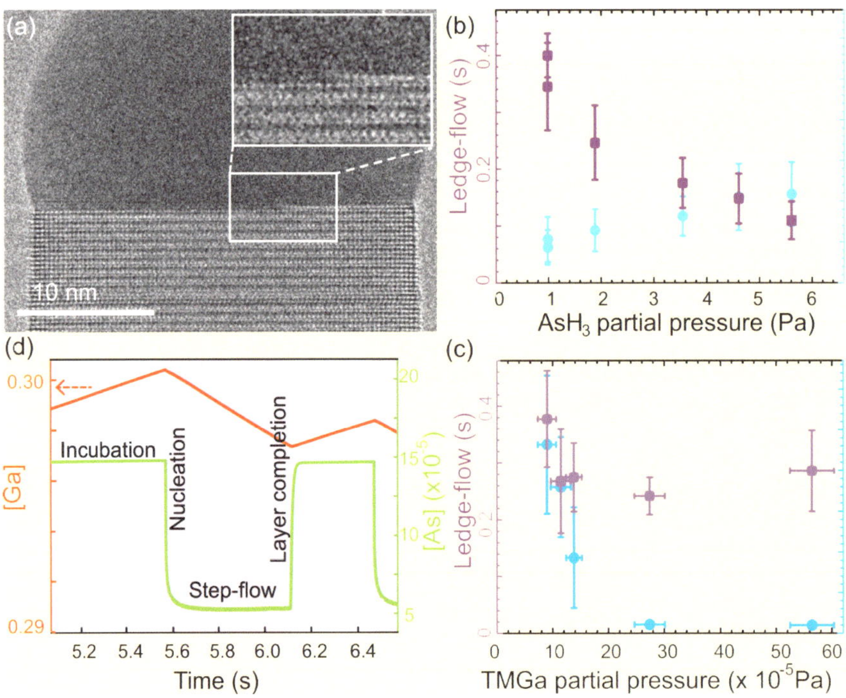

Figure 3.
(a) TEM image showing ledge-flow growth of an atomic bilayer. (b,c) Ledge-flow time as a function of As-precursor flow (b) and Ga-precursor flow (c). (d) A representative example of simulation of Ga and As concentrations in the catalyst. (Plots (b), (c) and (d) are adapted from Maliakkal et al. *[90] with permission. Further permissions should be directed to ACS.)*

During the incubation process the As concentration in the catalyst is in equilibrium with the ambient vapor. Once the layer nucleation happens the 'excess' As is consumed to form the GaAs nucleus and so As concentration quickly drops to a low level (where the As is in equilibrium with the solid GaAs nanowire). As soon as the layer is grown completely, the As % quickly rises and equilibrates again with the ambient vapor. Once this happens, the As contribution to the liquid supersaturation remains the same over the rest of the incubation period. However, the Ga building up in the catalyst keeps increasing the liquid chemical potential. Eventually, at some point after the liquid chemical potential is higher than the nucleation barrier, stochastically a nucleation event happens. Since the As % remains steady during the latter part of the incubation period, it is the Ga which is triggering nucleation of a new layer [90].

The study demonstrated independent control of layer nucleation (by Ga) and layer completion (by As) in GaAs nanowires growth [90]. The underlying reason for the nucleation of layer and ledge-flow to be controllable independently is the very low solubility of As and the high solubility of Ga in the Au catalyst. Several other III-V and II-VI compound semiconductors also consist of a nonmetallic species (group V or VI, e.g., N, O, P, S), and a metallic species (group II or III, e.g., Ga, In, Zn, Mg) [101]. These nonmetallic species typically dissolve very little in catalyst (gold or other typical transition metal catalysts) while the metallic species readily forms alloys. In cases were the amount of nonmetallic species collected in the catalyst and available for growth is low, the layer growth process will be restricted by availability of this nonmetallic species. Thus, independent control of layer nucleation and growth would be possible in several other nanowire systems too [90].

The occasions where controlling the layer nucleation and growth are extremely relevant could include doping and growth of ternary compounds. In VLS growth, the nucleation stage determines the crystal stacking of the entire atomic layer [1]. However, dopant/impurity incorporation happening would strongly depend on ledge-flow. Since impurity incorporation could be happening due to step trapping, a slow ledge-flow would help limit the impurity incorporation. On the contrary, for higher dopant incorporation, a fast ledge-flow could be advantageous [90].

8. Polytypism in III-V nanowires

Layer nucleation and growth for the different polytypes have been studied by in situ TEM. Before we discuss the key in situ results, let us discuss the concept of polytypism in nanowires and how the metastable structure could form. Nanowires enable the formation of metastable crystalline phases which do not form during its bulk growth. For example, most III-arsenides and III-phosphides form in the zincblende polytype when grown in bulk, because the bulk energy is lower for zincblende than wurtzite phase. But these materials can form in the wurtzite polytype in nanowires due to surface effects [2–4, 44, 110]. (Details of these crystal structures can be found elsewhere [4]). Controlled polytypism has great technological relevance because the electronic band structure depends on crystal structure. For example, GaP in the usual zincblende phase is an indirect bandgap material; while the wurtzite polytype has a pseudo-direct bandgap [111–113]. The valence and conduction bands of the two polytypes are often misaligned, so sections of one polytype in a matrix of the other polytype nanowire can confine electrons and/or holes. This enables crystal phase quantum dots with abrupt interfaces [114, 115]. Compositional quantum dots, on the other hand, often has a gradual variation of the composition (depending on the material combination chosen) deteriorating its properties.

Now let us first briefly discuss a simplified explanation for the occurrence of the metastable wurtzite structure in nanowires. During VLS growth, at appropriate catalyst contact angles, the nucleus of each layer is preferably formed at the triple-phase-line because it eliminates the energy cost of a preexisting liquid segment [1]. For nucleation happening at the triple-phase-line the nanowire surface energy is a key factor [1]. The surface energy of possible wurtzite side facets could be lower than the zincblende counterparts [1]. In such cases, the wurtzite structure can be more favorable than zincblende for energy minimization, depending on the catalyst supersaturation and relevant interface energies [1]. Extensive models proposed by several groups to correlate the observed crystal structure at different conditions can be found elsewhere [1, 96, 103, 104, 116–118].

In GaAs nanowire growth studied ex situ for a very wide range of V/III ratios it was found that very low V/III ratios gives zincblende, a higher V/III results in wurtzite, and an even higher V/III give zincblende again [110]. A high V/III ratio (i.e. higher AsH_3, which may also be interpreted as effectively lesser Ga) is associated with a smaller catalyst; while a low V/III results in bulged up catalyst with high contact angle [110]. There are also two intermediate transition regimes with mixed structures [110]. Theoretical models of Au-catalyzed GaAs growths could also simulate three different growth regimes [103, 104]. Often in typical experiment series a narrower range of V/III is studied, and thus it may happen that only a zincblende to wurtzite transition or only a wurtzite to zincblende transition is observed on increasing V/III ratio.

Jacobsson *et al.* observed two growth regimes in situ — at moderate V/III ratio wurtzite segments grew, and at low V/III the catalyst bulges and zincblende segments grew [84]. The wurtzite growth occurred while the nanowire-catalyst interface was one flat plane and the ledge-flow growth was gradual [84]. During zincblende growth the interface showed an oscillating truncated corner; the ledge-flow was "too rapid to observe" but was correlated to the cyclic dynamics of the truncation [84]. According to the study zincblende phase grows if any edge is truncated, whereas wurtzite grows only when the interface is flat [84]. They speculated that the truncation happening in the low V/III regime would make nucleation occur away from the triple-phase-line which in turn makes zincblende the preferred structure. (Note the authors had not claimed an if-and-only-if condition between interface geometry and crystal structure. However, in my personal experience, some readers misinterpreted that truncation was a necessity for low V/III zincblende growth.) The droplet contact angle was the key parameter in deciding the crystal structure [84]. Au:Ga ratio in the droplet was found to be not critical, hence a similar crystal structure–geometry correlation was speculated to be applicable to self-catalyzed wires too [84].

Polytypism in self-catalyzed GaAs nanowires was recently studied using in situ TEM-MBE by Panciera *et al* [99]. In this work the three regimes were observed as shown in **Figure 4e**, **g** and **h** (zincblende at low V/III ratio, wurtzite at higher V/III ratio, zincblende at even higher V/III ratio) [99]. In the high V/III zincblende regime, (which was not experimentally achievable in Jacobsson *et al.*), the nucleation was found to occur at the triple-phase-line, the ledge-flow was slow, and no truncation was observed. Their observations consistent with Jacobsson *et al.* include (i) control of crystal structure by contact angle, (ii) gradual ledge-flow and flat interface during wurtzite growth, and (iii) truncation and rapid ledge-flow observed in the low V/III zincblende growth [99].

However, it is not necessary that the low V/III zincblende growth can occur only with truncation. The growth of zincblende with a bulged particle (high contact angle) at low V/III ratio, but without truncation, has also been reported (**Figure 4f**) [89]. In this study the V/III ratio was decreased to observe bulging of the particle,

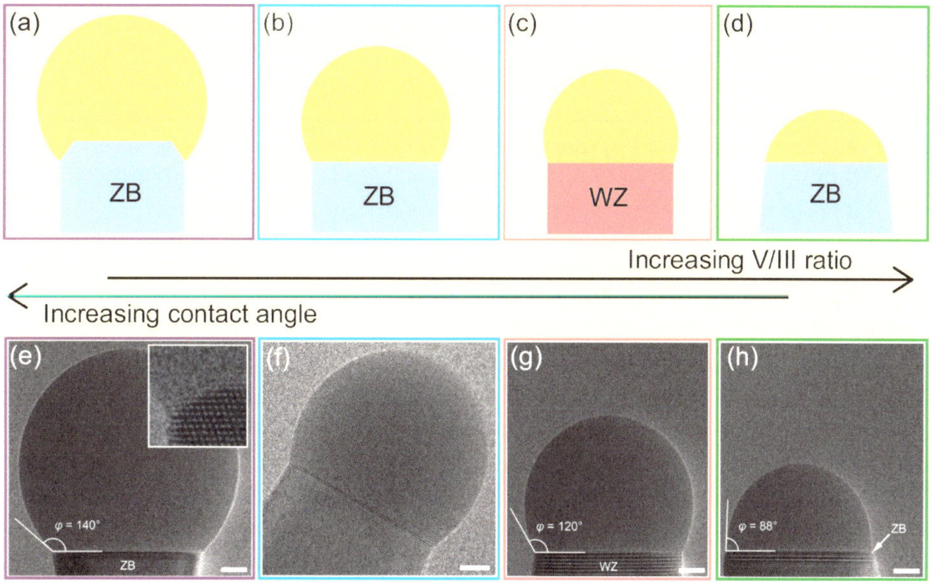

Figure 4.
(a)-(d) Schematic representation of crystal phases observed as a function of V/III ratio or catalyst contact angle. The catalyst-nanowire interface is either a single plane (b-d) or truncated (a). (e)-(h) shows TEM images with zincblende (ZB) or wurtzite (WZ) structures. Scalebars correspond to 5 nm. Inset of (e) shows a dynamic truncated corner during the zincblende growth at low V/III ratio. (e), (g) and (h) are adapted with permission from Panciera et al. *[99] Copyright (2020) American Chemical Society. Image (f) showing low V/III zincblende growth even without truncation is from another study – Maliakkal et al. [89]. Adapted with permission as per Creative Commons license.*

but a lower limit was set on the V/III ratio intentionally to avoid the truncation regime [89]. The V/III ratio was varied in small steps and maintained for some time to reach steady state [89] – this could be the reason why this intermediate zincblende regime with high contact angle and without truncation was observed in this study unlike the other two in situ studies discussed above [84, 99]. Thus, we can infer that the zincblende growth at low V/III does not necessarily require truncation. More detailed investigation is necessary to say if the nucleation happened at the triple-phase-line or the center in this case, and if the ledge-flow was gradual or instantaneous. Whether the presence of truncation makes the nucleation preferable away from the triple phase line is also an open question.

A heuristic explanation for the choice of crystal structure based on the currently available data [84, 89, 99] and theoretical calculations [1, 49] on GaAs VLS growth is as follows. The metastable wurtzite phase can grow only if the layer nucleates at the triple-phase-line [49]. If the nucleation is at triple-phase-line, either wurtzite or zincblende can form depending on the supersaturation and surface energies [1]. When nucleation occurs away from the triple-phase-line, it can form only zincblende structure [1, 49]. Now let us look at crystal structure as a function of the contact angle. Glas *et al.* predicted that nucleation is preferred at the triple-phase-line for a range of contact angles $[\pi - \beta_c; \beta_c]$, where the critical angle β_c is a function of the relevant interface energies [1, 49]. Thus, at intermediate contact angles (i.e. intermediate V/III ratios), nucleation occurs at the triple-phase-line and wurtzite structure is formed as observed(**Figure 4c,g**) [99]. For the lower contact angle (**Figure 4d,h**), nucleation was reported to be zincblende and at the triple-phase-line, [99] which demands that zincblende structure would have been the lower energy nucleus phase at those growth conditions [1]. For the higher contact angle regime, zincblende was found to grow, even without truncation [89]. It would have happened either (i) with nucleation away from the triple-phase-line giving zincblende structure or (ii) with nucleation at the triple-phase-line only, but with the zincblende structure having lower energy at those growth conditions. Note that in the above explanation or in references [1, 49] truncation was not explicitly needed to explain crystal phase switching. At extremely high contact angles and at extremely low contact angles, there could be either truncation or large tapering; [84, 99] but truncation is not a necessity for zincblende growth. The truncation might be responsible for the observed quasi-instantaneous ledge-flow though.

9. Open questions

Ideally, for an exact explanation of nanowire crystal phases at different conditions discussed above, one could theoretically model the system and compare the contact angles where the structure is predicted to switch phases and compare it with experimental values. Such calculations, and also models for other phenomena in nanowires, involve different interface energy terms. However, there are hardly any direct experimental measurements of interface energies at the different growth conditions even for common material systems. The solid surface energies (the solid-vapor interface energy to be more precise) would depend on the surface relaxations/reconstructions adapted by the system, which again depends on the growth condition [119]. There exists post-growth surface energy measurements on bulk materials, [120] but is inadequate for knowing nanowire surface energies during growth conditions. Some roundabout estimates have been made by comparing experimental observations with approximate models for finding surface energies during growth; [99, 108] having these values are certainly better than having nothing, but we need better measurements. The reason for not having more direct measurements is simple – they are challenging to perform and observe at nanowire growth conditions. The surface tension of Au-Si liquid catalyst has been beautifully measured by studying electric field-induced deformation [85]. This method can be used to study other material systems as well. We have to come up with smart strategies for measuring solid–liquid and solid-vapor interface energies.

A seemingly basic, but still ambiguous topic is what are the key parameters deciding the nanowire growth direction. For example, unless at very peculiar growth conditions, most III-V and II-VI nanowires grow in the <111>/<0001> B direction, [3, 4] even on amorphous substrates, [89, 113, 121] in fact even without a substrate [122]. (The <0001>B direction in wurtzite structure is equivalent to the <111>B of zincblende polytype.) A complete and accurate description in the VLS case would involve catalyst chemical potential, solid–liquid interface energy for different possible crystal planes in contact with the liquid catalyst, solid-vapor surface energy of nanowire sides, liquid–vapor interface energy, edge energies of the top facet, edge energies of the growing island, [49] edge energies at nanowire side corners, if or not a new layer has well-defined low-index facets or is the surface rather rounded, [123] effect of liquid ordering, [45] etc. These individual terms are a function of the growth parameters and catalyst composition. As mentioned in the previous paragraph, most of these values have not been measured yet. But all these factors present, perhaps there is some key factor(s) which overpowers at typical growth conditions?

A very interesting but unresolved question is the diffusion pathway of the reactants. If the group V or group VI species is expected to hardly dissolve in the catalyst during compound nanowire growth, is it necessary that it should diffuse through the volume of the catalyst? Could these species be diffusing through the catalyst-nanowire surface instead? With the current technology it is not possible to watch the trajectory of each individual atom. However, perhaps there could be some strategic in situ experiments, which in combination with appropriate rigorous theoretical simulations, can solve this puzzle. The radius dependence of the ledge-flow time might distinguish if the diffusion is through bulk or interface. The dynamics of the ledge-flow and the shape of the growing layer might also serve as a tool. That said, it is not necessary that there be a unique answer to this puzzle even for a given catalyst-nanowire system and catalyst phase (i.e. VLS or VSS), perhaps it could be dependent on the growth conditions. Another approach to this puzzle could be – diffusion need not even be the rate limiting process; in such a case why care about it. But this is nonetheless an interesting unanswered riddle, where in situ TEM can be extremely valuable.

10. Summary

Several in situ techniques, including in situ TEM, has been used to study nanowire growth. In situ TEM studies revealed that the growth dynamics of compound nanowires (e.g. III-V nanowires like GaAs) is fundamentally different from elemental nanowires. This can be understood by the difference in solubility of the nanowire species in the catalyst, which was also investigated by in situ TEM. Due to this concentration difference the layer nucleation and layer completion processes could be independently controlled. The growth dynamics has been studied in relation with the crystal structure and nanowire-catalyst interface morphology.

Author details

Carina B. Maliakkal
Centre for Analysis and Synthesis, Lund University, Lund, Sweden

*Address all correspondence to: carina_babu.maliakkal@chem.lu.se, carinab.maliakkal@gmail.com

References

[1] Glas F, Harmand J-C, Patriarche G. Why Does Wurtzite Form in Nanowires of III-V Zinc Blende Semiconductors? Phys Rev Lett. 2007 Oct 5;99(14):146101.

[2] Persson AI, Larsson MW, Stenström S, Ohlsson BJ, Samuelson L, Wallenberg LR. Solid-phase diffusion mechanism for GaAs nanowire growth. Nature Materials. 2004 Oct;3(10):677-681.

[3] Joyce HJ, Wong-Leung J, Gao Q, Tan HH, Jagadish C. Phase Perfection in Zinc Blende and Wurtzite III−V Nanowires Using Basic Growth Parameters. Nano Lett. 2010 Mar 10;10(3):908-915.

[4] Dick KA, Caroff P, Bolinsson J, Messing ME, Johansson J, Deppert K, et al. Control of III–V nanowire crystal structure by growth parameter tuning. Semicond Sci Technol. 2010;25(2):024009.

[5] Hu J, Ouyang M, Yang P, Lieber CM. Controlled growth and electrical properties of heterojunctions of carbon nanotubes and silicon nanowires. Nature. 1999 May;399(6731):48-51.

[6] Guo YN, Zou J, Paladugu M, Wang H, Gao Q, Tan HH, et al. Structural characteristics of GaSb/GaAs nanowire heterostructures grown by metal-organic chemical vapor deposition. Appl Phys Lett. 2006 Dec 4;89(23):231917.

[7] Caroff P, Messing ME, Borg BM, Dick KA, Deppert K, Wernersson L-E. InSb heterostructure nanowires: MOVPE growth under extreme lattice mismatch. Nanotechnology. 2009 Nov;20(49):495606.

[8] Barth S, S. Seifner M, Bernardi J. Microwave-assisted solution–liquid–solid growth of Ge 1−x Sn x nanowires with high tin content. Chemical Communications. 2015;51(61):12282-12285.

[9] Güniat L, Caroff P, Fontcuberta i Morral A. Vapor Phase Growth of Semiconductor Nanowires: Key Developments and Open Questions. Chem Rev. 2019 Aug 14;119(15):8958-8971.

[10] Hsueh T-J, Hsu C-L, Chang S-J, Guo P-W, Hsieh J-H, Chen I-C. Cu2O/n-ZnO nanowire solar cells on ZnO:Ga/glass templates. Scripta Materialia. 2007 Jul 1;57(1):53-56.

[11] Krogstrup P, Jørgensen HI, Heiss M, Demichel O, Holm JV, Aagesen M, et al. Single-nanowire solar cells beyond the Shockley–Queisser limit. Nature Photonics. 2013 Apr;7(4):306-310.

[12] Wallentin J, Anttu N, Asoli D, Huffman M, Åberg I, Magnusson MH, et al. InP Nanowire Array Solar Cells Achieving 13.8% Efficiency by Exceeding the Ray Optics Limit. Science. 2013 Mar 1;339(6123):1057-1060.

[13] Otnes G, Borgström MT. Towards high efficiency nanowire solar cells. Nano Today. 2017 Feb 1;12:31-45.

[14] Cui Y, Duan X, Hu J, Lieber CM. Doping and Electrical Transport in Silicon Nanowires. J Phys Chem B. 2000 Jun 1;104(22):5213-5216.

[15] Huang Y, Duan X, Cui Y, Lauhon LJ, Kim K-H, Lieber CM. Logic Gates and Computation from Assembled Nanowire Building Blocks. Science. 2001 Nov 9;294(5545):1313-1317.

[16] Li Y, Qian F, Xiang J, Lieber CM. Nanowire electronic and optoelectronic devices. Materials Today. 2006 Oct 1;9(10):18-27.

[17] Könenkamp R, Word RC, Schlegel C. Vertical nanowire light-emitting diode. Appl Phys Lett. 2004 Dec 9;85(24):6004-6006.

[18] Johnson JC, Choi H-J, Knutsen KP, Schaller RD, Yang P, Saykally RJ. Single gallium nitride nanowire lasers. Nature Materials. 2002 Oct;1(2):106-110.

[19] Huang MH, Mao S, Feick H, Yan H, Wu Y, Kind H, et al. Room-Temperature Ultraviolet Nanowire Nanolasers. Science. 2001 Jun 8;292(5523):1897-1899.

[20] Wang J, Gudiksen MS, Duan X, Cui Y, Lieber CM. Highly Polarized Photoluminescence and Photodetection from Single Indium Phosphide Nanowires. Science. 2001 Aug 24;293(5534):1455-1457.

[21] Yan C, Wang J, Wang X, Kang W, Cui M, Foo CY, et al. An Intrinsically Stretchable Nanowire Photodetector with a Fully Embedded Structure. Advanced Materials. 2014;26(6):943-950.

[22] Hochbaum AI, Chen R, Delgado RD, Liang W, Garnett EC, Najarian M, et al. Enhanced thermoelectric performance of rough silicon nanowires. Nature. 2008 Jan;451(7175):163-167.

[23] Wei Q, Lieber CM. Synthesis of Single Crystal Bismuth-Telluride and Lead-Telluride Nanowires for New Thermoelectric Materials. MRS Online Proceedings Library Archive. 1999 ed;581.

[24] Cui Y, Wei Q, Park H, Lieber CM. Nanowire Nanosensors for Highly Sensitive and Selective Detection of Biological and Chemical Species. Science. 2001 Aug 17;293(5533):1289-1292.

[25] Zhao Y, You SS, Zhang A, Lee J-H, Huang J, Lieber CM. Scalable ultrasmall three-dimensional nanowire transistor probes for intracellular recording. Nature Nanotechnology. 2019 Aug;14(8):783-790.

[26] Hazard TM, Gyenis A, Di Paolo A, Asfaw AT, Lyon SA, Blais A, et al. Nanowire Superinductance Fluxonium Qubit. Phys Rev Lett. 2019 Jan 10;122(1):010504.

[27] Ku J, Manucharyan V, Bezryadin A. Superconducting nanowires as nonlinear inductive elements for qubits. Phys Rev B. 2010 Oct 13;82(13):134518.

[28] Ek M, Filler MA. Atomic-Scale Choreography of Vapor–Liquid–Solid Nanowire Growth. Acc Chem Res. 2018 Jan 16;51(1):118-126.

[29] Tian B, Xie P, Kempa TJ, Bell DC, Lieber CM. Single-crystalline kinked semiconductor nanowire superstructures. Nature Nanotechnology. 2009 Dec;4(12):824-829.

[30] Hayden O, Agarwal R, Lu W. Semiconductor nanowire devices. Nano Today. 2008 Oct 1;3(5):12-22.

[31] Riveros G, Gómez H, Schrebler R, Marotti RE, Dalchiele EA. An In Situ EIS Study during the Electrochemical Growth of Copper Nanowires into Porous Polycarbonate Membranes. Electrochem Solid-State Lett. 2007 Dec 27;11(3):K19.

[32] Khan MI, Penchev M, Jing X, Wang X, Bozhilov KN, Ozkan M, et al. Electrochemical Growth of InSb Nanowires and Report of a Single Nanowire Field Effect Transistor. Journal of Nanoelectronics and Optoelectronics. 2008 Jul 1;3(2):199-202.

[33] Fahrenkrug E, Gu J, Jeon S, Veneman PA, Goldman RS, Maldonado S. Room-Temperature Epitaxial Electrodeposition of Single-Crystalline Germanium Nanowires

at the Wafer Scale from an Aqueous Solution. Nano Lett. 2014 Feb 12;14(2):847-852.

[34] Cheek Q, Fahrenkrug E, Hlynchuk S, Alsem DH, Salmon NJ, Maldonado S. In Situ Transmission Electron Microscopy Measurements of Ge Nanowire Synthesis with Liquid Metal Nanodroplets in Water. ACS Nano. 2020 Mar 24;14(3):2869-2879.

[35] Wagner RS, Ellis WC. Vapor-liquid-solid mechanism of single crystal growth. Appl Phys Lett. 1964 Mar 1;4(5):89-90.

[36] Bootsma GA, Gassen HJ. A quantitative study on the growth of silicon whiskers from silane and germanium whiskers from germane. Journal of Crystal Growth. 1971 Aug 1;10(3):223-234.

[37] Schouler MC, Cheynet MC, Sestier K, Garden J, Gadelle P. New filamentous deposits in the boron-carbon system. Carbon. 1997 Jan 1;35(7):993-1000.

[38] Mandl B, Stangl J, Hilner E, Zakharov AA, Hillerich K, Dey AW, et al. Growth Mechanism of Self-Catalyzed Group III–V Nanowires. Nano Lett. 2010 Nov 10;10(11):4443-4449.

[39] Ramdani MR, Harmand JC, Glas F, Patriarche G, Travers L. Arsenic Pathways in Self-Catalyzed Growth of GaAs Nanowires. Crystal Growth & Design. 2013 Jan 2;13(1):91-96.

[40] Heon Kim Y, Woo Park D, Jun Lee S. Gallium-droplet behaviors of self-catalyzed GaAs nanowires: A transmission electron microscopy study. Appl Phys Lett. 2012 Jan 16;100(3):033117.

[41] Zhang Z, Wang Y, Li H, Yuan W, Zhang X, Sun C, et al. Atomic-Scale Observation of Vapor–Solid Nanowire Growth via Oscillatory Mass Transport. ACS Nano. 2016 Jan 26;10(1):763-769.

[42] Lee SC, Dawson LR, Brueck SRJ, Jiang Y-B. Anisotropy of selective epitaxy in nanoscale-patterned growth: GaAs nanowires selectively grown on a SiO2-patterned (001) substrate by molecular-beam epitaxy. Journal of Applied Physics. 2005 Dec 1;98(11):114312.

[43] Noborisaka J, Motohisa J, Hara S, Fukui T. Fabrication and characterization of freestanding GaAs/AlGaAs core-shell nanowires and AlGaAs nanotubes by using selective-area metalorganic vapor phase epitaxy. Appl Phys Lett. 2005 Aug 24;87(9):093109.

[44] Jacobsson D, Lehmann S, Dick KA. Zincblende-to-wurtzite interface improvement by group III loading in Au-seeded GaAs nanowires. Phys Status Solidi RRL. 2013 Oct 1;7(10):855-859.

[45] Oh SH, Chisholm MF, Kauffmann Y, Kaplan WD, Luo W, Rühle M, et al. Oscillatory Mass Transport in Vapor-Liquid-Solid Growth of Sapphire Nanowires. Science. 2010 Oct 22;330(6003):489-493.

[46] Wen C-Y, Tersoff J, Hillerich K, Reuter MC, Park JH, Kodambaka S, et al. Periodically Changing Morphology of the Growth Interface in Si, Ge, and GaP Nanowires. Phys Rev Lett. 2011 Jul 6;107(2):025503.

[47] Gamalski AD, Ducati C, Hofmann S. Cyclic Supersaturation and Triple Phase Boundary Dynamics in Germanium Nanowire Growth. J Phys Chem C. 2011 Mar 24;115(11):4413-4417.

[48] Tornberg M, Maliakkal CB, Jacobsson D, Dick KA, Johansson J. Limits of III–V Nanowire Growth Based on Droplet Dynamics. J Phys Chem Lett. 2020 Apr 16;11(8):2949-2954.

[49] Harmand J-C, Patriarche G, Glas F, Panciera F, Florea I, Maurice J-L, et al. Atomic Step Flow on a Nanofacet. Phys Rev Lett. 2018 Oct 19;121(16):166101.

[50] Tchernycheva M, Harmand JC, Patriarche G, Travers L, Cirlin GE. Temperature conditions for GaAs nanowire formation by Au-assisted molecular beam epitaxy. Nanotechnology. 2006 Jul;17(16):4025-4030.

[51] Jo J, Tchoe Y, Yi G-C, Kim M. Real-Time Characterization Using in situ RHEED Transmission Mode and TEM for Investigation of the Growth Behaviour of Nanomaterials. Scientific Reports. 2018 Jan 26;8(1):1694.

[52] Heurlin M, Anttu N, Camus C, Samuelson L, Borgström MT. In Situ Characterization of Nanowire Dimensions and Growth Dynamics by Optical Reflectance. Nano Lett. 2015 May 13;15(5):3597-3602.

[53] Braun MR, Güniat L, Morral AFI, McIntyre PC. In-situ reflectometry to monitor locally-catalyzed initiation and growth of nanowire assemblies. Nanotechnology. 2020 Jun;31(33):335703.

[54] Schroth P, Köhl M, Hornung J-W, Dimakis E, Somaschini C, Geelhaar L, et al. Evolution of Polytypism in GaAs Nanowires during Growth Revealed by Time-Resolved in situ x-ray Diffraction. Phys Rev Lett. 2015 Feb 5;114(5):055504.

[55] Krogstrup P, Hannibal Madsen M, Hu W, Kozu M, Nakata Y, Nygård J, et al. In-situ x-ray characterization of wurtzite formation in GaAs nanowires. Appl Phys Lett. 2012 Feb 27;100(9):093103.

[56] Kirkham M, Wang ZL, Snyder RL. Tracking the catalyzed growth process of nanowires by in situ x-ray diffraction. Journal of Applied Physics. 2010 Jul 1;108(1):014304.

[57] Sivaram SV, Hui HY, de la Mata M, Arbiol J, Filler MA. Surface Hydrogen Enables Subeutectic Vapor–Liquid–Solid Semiconductor Nanowire Growth. Nano Lett. 2016 Nov 9;16(11):6717-6723.

[58] Sivaram SV, Shin N, Chou L-W, Filler MA. Direct Observation of Transient Surface Species during Ge Nanowire Growth and Their Influence on Growth Stability. J Am Chem Soc. 2015 Aug 12;137(31):9861-9869.

[59] Shin N, Chi M, Filler MA. Interplay between Defect Propagation and Surface Hydrogen in Silicon Nanowire Kinking Superstructures. ACS Nano. 2014 Apr 22;8(4):3829-3835.

[60] Shin N, Filler MA. Controlling Silicon Nanowire Growth Direction via Surface Chemistry. Nano Lett. 2012 Jun 13;12(6):2865-2870.

[61] Fernández-Garrido S, Zettler JK, Geelhaar L, Brandt O. Monitoring the Formation of Nanowires by Line-of-Sight Quadrupole Mass Spectrometry: A Comprehensive Description of the Temporal Evolution of GaN Nanowire Ensembles. Nano Lett. 2015 Mar 11;15(3):1930-1937.

[62] Volkov Y, Mitchell S, Gaponik N, Rakovich YP, Donegan JF, Kelleher D, et al. In-Situ Observation of Nanowire Growth from Luminescent CdTe Nanocrystals in a Phosphate Buffer Solution. ChemPhysChem. 2004 Oct 18;5(10):1600-1602.

[63] Kolíbal M, Pejchal T, Vystavěl T, Šikola T. The Synergic Effect of Atomic Hydrogen Adsorption and Catalyst Spreading on Ge Nanowire Growth Orientation and Kinking. Nano Lett. 2016 Aug 10;16(8):4880-4886.

[64] Prikhodko SV, Sitzman S, Gambin V, Kodambaka S. In situ

electron backscattered diffraction of individual GaAs nanowires. Ultramicroscopy. 2008 Dec 1;109(1):133-138.

[65] Kolíbal M, Vystavěl T, Varga P, Šikola T. Real-Time Observation of Collector Droplet Oscillations during Growth of Straight Nanowires. Nano Lett. 2014 Apr 9;14(4):1756-1761.

[66] Sun Y, Gao J, Zhu R, Xu J, Chen L, Zhang J, et al. In situ observation of ZnO nanowire growth on zinc film in environmental scanning electron microscope. J Chem Phys. 2010 Mar 28;132(12):124705.

[67] Huang X, Wang Z-J, Weinberg G, Meng X-M, Willinger M-G. In Situ Scanning Electron Microscopy Observation of Growth Kinetics and Catalyst Splitting in Vapor–Liquid–Solid Growth of Nanowires. Advanced Functional Materials. 2015 Oct 1;25(37):5979-5987.

[68] Wu Y, Yang P. Direct Observation of Vapor–Liquid–Solid Nanowire Growth. J Am Chem Soc. 2001 Apr 1;123(13):3165-3166.

[69] Ross FM, Tersoff J, Reuter MC. Sawtooth Faceting in Silicon Nanowires. Phys Rev Lett. 2005 Sep 29;95(14):146104.

[70] Kodambaka S, Tersoff J, Reuter MC, Ross FM. Diameter-Independent Kinetics in the Vapor-Liquid-Solid Growth of Si Nanowires. Phys Rev Lett. 2006 Mar 9;96(9):096105.

[71] Hannon JB, Kodambaka S, Ross FM, Tromp RM. The influence of the surface migration of gold on the growth of silicon nanowires. Nature. 2006 Mar;440(7080):69-71.

[72] Kodambaka S, Tersoff J, Reuter MC, Ross FM. Germanium Nanowire Growth Below the Eutectic Temperature. Science. 2007 May 4;316(5825):729-732.

[73] Hofmann S, Sharma R, Wirth CT, Cervantes-Sodi F, Ducati C, Kasama T, et al. Ledge-flow-controlled catalyst interface dynamics during Si nanowire growth. Nature Materials. 2008 May;7(5):372-375.

[74] Wen C-Y, Reuter MC, Bruley J, Tersoff J, Kodambaka S, Stach EA, et al. Formation of Compositionally Abrupt Axial Heterojunctions in Silicon-Germanium Nanowires. Science. 2009 Nov 27;326(5957):1247-1250.

[75] Madras P, Dailey E, Drucker J. Kinetically Induced Kinking of Vapor–Liquid–Solid Grown Epitaxial Si Nanowires. Nano Lett. 2009 Nov 11;9(11):3826-3830.

[76] Kallesøe C, Wen C-Y, Mølhave K, Bøggild P, Ross FM. Measurement of Local Si-Nanowire Growth Kinetics Using In situ Transmission Electron Microscopy of Heated Cantilevers. Small. 2010 Sep 20;6(18):2058-2064.

[77] Wen C-Y, Reuter MC, Tersoff J, Stach EA, Ross FM. Structure, Growth Kinetics, and Ledge Flow during Vapor–Solid–Solid Growth of Copper-Catalyzed Silicon Nanowires. Nano Lett. 2010 Feb 10;10(2):514-519.

[78] Wen C-Y, Tersoff J, Reuter MC, Stach EA, Ross FM. Step-Flow Kinetics in Nanowire Growth. Phys Rev Lett. 2010 Nov 5;105(19):195502.

[79] Diaz RE, Sharma R, Jarvis K, Zhang Q, Mahajan S. Direct observation of nucleation and early stages of growth of GaN nanowires. Journal of Crystal Growth. 2012 Feb 15;341(1):1-6.

[80] Chou Y-C, Hillerich K, Tersoff J, Reuter MC, Dick KA, Ross FM. Atomic-Scale Variability and Control of III-V Nanowire Growth Kinetics. Science. 2014 Jan 17;343(6168):281-284.

[81] Schwarz KW, Tersoff J, Kodambaka S, Ross FM.

Jumping-Catalyst Dynamics in Nanowire Growth. Phys Rev Lett. 2014 Jul 30;113(5):055501.

[82] Chou Y-C, Panciera F, C. Reuter M, A. Stach E, M. Ross F. Nanowire growth kinetics in aberration corrected environmental transmission electron microscopy. Chemical Communications. 2016;52(33):5686-5689.

[83] Gamalski AD, Tersoff J, Stach EA. Atomic Resolution in Situ Imaging of a Double-Bilayer Multistep Growth Mode in Gallium Nitride Nanowires. Nano Lett. 2016 Apr 13;16(4):2283-2288.

[84] Jacobsson D, Panciera F, Tersoff J, Reuter MC, Lehmann S, Hofmann S, et al. Interface dynamics and crystal phase switching in GaAs nanowires. Nature. 2016 Mar 16;531(7594):317.

[85] Panciera F, Norton MM, Alam SB, Hofmann S, Mølhave K, Ross FM. Controlling nanowire growth through electric field-induced deformation of the catalyst droplet. Nature Communications. 2016 Jul 29;7:12271.

[86] Panciera F, Tersoff J, Gamalski AD, Reuter MC, Zakharov D, Stach EA, et al. Surface Crystallization of Liquid Au–Si and Its Impact on Catalysis. Advanced Materials. 2019;31(5):1806544.

[87] Kim BJ, Tersoff J, Kodambaka S, Reuter MC, Stach EA, Ross FM. Kinetics of Individual Nucleation Events Observed in Nanoscale Vapor-Liquid-Solid Growth. Science. 2008 Nov 14;322(5904):1070-1073.

[88] Kim BJ, Wen C-Y, Tersoff J, Reuter MC, Stach EA, Ross FM. Growth Pathways in Ultralow Temperature Ge Nucleation from Au. Nano Lett. 2012 Nov 14;12(11):5867-5872.

[89] Maliakkal CB, Jacobsson D, Tornberg M, Persson AR, Johansson J, Wallenberg R, et al. In situ analysis of catalyst composition during gold catalyzed GaAs nanowire growth. Nat Commun. 2019 Oct 8;10(1):1-9.

[90] Maliakkal CB, Mårtensson EK, Tornberg MU, Jacobsson D, Persson AR, Johansson J, et al. Independent Control of Nucleation and Layer Growth in Nanowires. ACS Nano. 2020 Feb 12;14(4):3868-3875.

[91] Lenrick F, Ek M, Deppert K, Samuelson L, Wallenberg LR. Straight and kinked InAs nanowire growth observed in situ by transmission electron microscopy. Nano Res. 2014 Aug 1;7(8):1188-1194.

[92] Song M, Lee J, Wang B, A. Legg B, Hu S, Chun J, et al. In situ characterization of kinetics and mass transport of PbSe nanowire growth via LS and VLS mechanisms. Nanoscale. 2019;11(13):5874-5878.

[93] Mølhave K, Wacaser BA, Petersen DH, Wagner JB, Samuelson L, Bøggild P. Epitaxial Integration of Nanowires in Microsystems by Local Micrometer-Scale Vapor-Phase Epitaxy. Small. 2008;4(10):1741-1746.

[94] Kodambaka S, Hannon JB, Tromp RM, Ross FM. Control of Si Nanowire Growth by Oxygen. Nano Lett. 2006 Jun 1;6(6):1292-1296.

[95] Glas F, Harmand J-C, Patriarche G. Nucleation Antibunching in Catalyst-Assisted Nanowire Growth. Phys Rev Lett. 2010 Mar 31;104(13):135501.

[96] Dubrovskii VG, Sibirev NV, Harmand JC, Glas F. Growth kinetics and crystal structure of semiconductor nanowires. Phys Rev B. 2008 Dec 1;78(23):235301.

[97] Dubrovskii VG, Sokolova ZhV, Rylkova MV, Zhiglinsky AA. Composition and contact angle of Au-III-V droplets on top of Au-catalyzed III-V nanowires. Materials Physics & Mechanics. 2018 Mar;36(1):1-7.

[98] Dubrovskii VG, Sibirev NV. Growth rate of a crystal facet of arbitrary size and growth kinetics of vertical nanowires. Phys Rev E. 2004 Sep 15;70(3):031604.

[99] Panciera F, Baraissov Z, Patriarche G, Dubrovskii VG, Glas F, Travers L, et al. Phase Selection in Self-catalyzed GaAs Nanowires. Nano Lett. 2020 Mar 11;20(3):1669-1675.

[100] Stach EA, Pauzauskie PJ, Kuykendall T, Goldberger J, He R, Yang P. Watching GaN Nanowires Grow. Nano Lett. 2003 Jun 1;3(6):867-869.

[101] Massalski TB. Binary alloy phase diagrams. II. Materials Park, Ohio: American Society for Metals; 1990.

[102] Glas F, Ramdani MR, Patriarche G, Harmand J-C. Predictive modeling of self-catalyzed III-V nanowire growth. Phys Rev B. 2013 Nov 8;88(19):195304.

[103] Mårtensson EK, Lehmann S, Dick KA, Johansson J. Simulation of GaAs Nanowire Growth and Crystal Structure. Nano Lett. 2019 Feb 13;19(2):1197-1203.

[104] Dubrovskii VG. Influence of the group V element on the chemical potential and crystal structure of Au-catalyzed III-V nanowires. Appl Phys Lett. 2014 Feb 3;104(5):053110.

[105] Dubrovskii VG, Grecenkov J. Zeldovich Nucleation Rate, Self-Consistency Renormalization, and Crystal Phase of Au-Catalyzed GaAs Nanowires. Crystal Growth & Design. 2015 Jan 7;15(1):340-347.

[106] D. Leshchenko E, Ghasemi M, G. Dubrovskii V, Johansson J. Nucleation-limited composition of ternary III–V nanowires forming from quaternary gold based liquid alloys. CrystEngComm. 2018;20(12):1649-1655.

[107] Harmand JC, Patriarche G, Péré-Laperne N, Mérat-Combes M-N, Travers L, Glas F. Analysis of vapor-liquid-solid mechanism in Au-assisted GaAs nanowire growth. Appl Phys Lett. 2005 Nov 7;87(20):203101.

[108] Tornberg M, Jacobsson D, Persson AR, Wallenberg R, Dick KA, Kodambaka S. Kinetics of Au–Ga Droplet Mediated Decomposition of GaAs Nanowires. Nano Lett. 2019 Jun 12;19(6):3498-3504.

[109] Prince A, Raynor GV, Evans DS, Institute of Metals. Phase diagrams of ternary gold alloys. London; Brookfield, VT: Institute of Metals; 1990.

[110] Lehmann S, Jacobsson D, Dick KA. Crystal phase control in GaAs nanowires: opposing trends in the Ga- and As-limited growth regimes. Nanotechnology. 2015;26(30):301001.

[111] Gagliano L, Belabbes A, Albani M, Assali S, Verheijen MA, Miglio L, et al. Pseudodirect to Direct Compositional Crossover in Wurtzite GaP/InxGa1–xP Core–Shell Nanowires. Nano Lett. 2016 Dec 14;16(12):7930-7936.

[112] Assali S, Greil J, Zardo I, Belabbes A, de Moor MWA, Koelling S, et al. Optical study of the band structure of wurtzite GaP nanowires. Journal of Applied Physics. 2016 Jul 25;120(4):044304.

[113] Maliakkal CB, Gokhale M, Parmar J, Bapat RD, Chalke BA, Ghosh S, et al. Growth, structural and optical characterization of wurtzite GaP nanowires. Nanotechnology. 2019 Apr;30(25):254002.

[114] Dick KA, Thelander C, Samuelson L, Caroff P. Crystal Phase Engineering in Single InAs Nanowires. Nano Lett. 2010 Sep 8;10(9):3494-3499.

[115] Akopian N, Patriarche G, Liu L, Harmand J-C, Zwiller V. Crystal Phase

Quantum Dots. Nano Lett. 2010 Apr 14;10(4):1198-1201.

[116] Akiyama T, Sano K, Nakamura K, Ito T. An Empirical Potential Approach to Wurtzite–Zinc-Blende Polytypism in Group III–V Semiconductor Nanowires. Jpn J Appl Phys. 2006 Feb 24;45(3L):L275.

[117] Pankoke V, Kratzer P, Sakong S. Calculation of the diameter-dependent polytypism in GaAs nanowires from an atomic motif expansion of the formation energy. Phys Rev B. 2011 Aug 10;84(7):075455.

[118] Johansson J, Bolinsson J, Ek M, Caroff P, Dick KA. Combinatorial Approaches to Understanding Polytypism in III–V Nanowires. ACS Nano. 2012 Jul 24;6(7):6142-6149.

[119] Moll N, Kley A, Pehlke E, Scheffler M. GaAs equilibrium crystal shape from first principles. Phys Rev B. 1996 Sep 15;54(12):8844-8855.

[120] Messmer C, Bilello JC. The surface energy of Si, GaAs, and GaP. Journal of Applied Physics. 1981 Jul 1;52(7):4623-4629.

[121] Hui TA, Wang F, Han N, Yip S, Xiu F, J. Hou J, et al. High-performance indium phosphide nanowires synthesized on amorphous substrates: from formation mechanism to optical and electrical transport measurements. Journal of Materials Chemistry. 2012;22(21):10704-10708.

[122] Heurlin M, Magnusson MH, Lindgren D, Ek M, Wallenberg LR, Deppert K, et al. Continuous gas-phase synthesis of nanowires with tunable properties. Nature. 2012 Dec;492(7427):90-94.

[123] Jiang N, Wong-Leung J, Joyce HJ, Gao Q, Tan HH, Jagadish C. Understanding the True Shape of Au-Catalyzed GaAs Nanowires. Nano Lett. 2014 Oct 8;14(10):5865-5872.

Section 3

Other Nanowires

Chapter 7

Indium (In)-Catalyzed Silicon Nanowires (Si NWs) Grown by the Vapor–Liquid–Solid (VLS) Mode for Nanoscale Device Applications

M. Ajmal Khan and Yasuaki Ishikawa

Abstract

Stacking fault free and planar defects (twin plane) free catalyzed Si nanowires (Si NWs) is essential for the carrier transport in the nanoscale devices applications. In this chapter, In-catalyzed, vertically aligned and cone-shaped Si NWs arrays were grown by using vapor–liquid–solid (VLS) mode on Si (111) substrates. We have successfully controlled the verticality and (111)-orientation of Si NWs as well as scaled down the diameter to 18 nm. The density of Si NWs was also enhanced from 2.5 μm^{-2} to 70 μm^{-2}. Such vertically aligned, (111)-oriented p-type Si NWs are very important for the nanoscale device applications including Si NWs/c-Si tandem solar cells and p-Si NWs/n-InGaZnO Heterojunction LEDs. Next, the influence of substrate growth temperature (T_S), cooling rate ($\Delta T_S/\Delta t$) on the formation of planar defects, twining along [112] direction and stacking fault in Si NWs perpendicular to (111)-orientation were deeply investigated. Finally, one simple model was proposed to explain the formation of stacking fault, twining of planar defects in perpendicular direction to the axial growth direction of Si NWs. When the T_S was decreased from 600°C with the cooling rate of 100°C/240 sec to room temperature (RT) after Si NWs growth then the twin planar defects perpendicular to the substrate and along different segments of (111)-oriented Si NWs were observed.

Keywords: silicon nanowires (Si NWs), VLS growth mode, contact angle, vertically aligned, In-catalyst, twining plane defects, stacking fault, nanoscale devices, solar cells

1. Introduction

Recently much interest has been developed to control the band gap as a function of diameter of Si nanowires (Si NWs) to exploit the quantum size effect for photovoltaic applications [1–5], and its extension according to Moore's law in view of the ongoing downscaling of integrated circuits (ICs) technologies as well as nano devices. Specially Si NWs are remarkably important for the fabrication of nanoscale devices such as transistors [6], sensors [7, 8], and thermoelectric devices [9]. Vapor–liquid–solid (VLS) mechanism provides a unique opportunity to investigate the crystalline quality and structures of single NW, where density, orientation, and periodicity of Si NWs system can be influenced by growth parameters such as temperature, pressure, plasma treatment, dopants [10–12], and the type of catalyst [13, 14], surface condition of substrate as well as size [15] of the metal Nano-droplets (NDs), shown in **Figure 1**. Si NWs growth by VLS mode using various material catalysts, such as Au, Al, Ga, In, Pb, Sn and Zn have been reported [16–25].

Many researchers already grown vertically aligned Si NWs using Au-catalyst, which is not useful candidate for the application of nanoscale devices including solar cell and LEDs because it creates deep acceptor energy level at 0.54 eV in the Si band gap, whereas In-catalyst creates shallow accepter energy level at 0.16 eV in the Si band gap. Au-catalyst particles are strongly degrading the minority carrier life time, while In-catalyst particles are boosting to the carrier life time [26]. Previously, randomly oriented Si NWs were grown by Jeon and Kamisako *et al.* using different type of catalysts including In-metal in the VLS growth mode [27–29]. Usually, VLS grown Si NWs system shows complex faceting [30] including hexagonal structure [31]. Ordered arrays of planar faults were reported by Ohno *et al.* which resulted in new phases and properties of well-known materials [32]. Several researchers reported about the twinning and generation of polytype defects and their control in III-V materials based NWs [11, 12]. Generation of polytype defects in group IV-semiconductors were less explored [13, 33]. In Si NWs system {111} planar faults were confirmed along the growth axis of ⟨112⟩-orientation [14]. Such twin planar faults along a ⟨111⟩ direction, are considered as a stacking faults in the ABC stacking sequence. Ultimately, this arrangement gives rise to some local hexagonal ordering, for example, ABA, and leading to polytypes. It was also observed that isolated defects of these {111} faults of NWs can trap to the Au (gold)-catalyst atoms [34]. Such arrangements and characteristics have significant influence on the impurity distributions, electrical and optical properties of the Si NWs based nano-devices.

The defects study like twin planar defects as well as stacking fault were not rigorously investigated in the case of vertically aligned In-catalyzed (111)-oriented Si NWs. However, relatively very few investigations have been made about the stacking fault and twin planar defects in In-catalyzed Si NWs grown by VLS growth [26]. Zhan *et al.* reported the numerical study of Si NWs which discuss about the perpendicularly aligned stacking fault layers, the extrinsic stacking fault (eSF) and 9-rhombohedral (9R)-polytype [35]. Large reduction of thermal conductivity in Si NWs was induced by extrinsic stacking fault (eSF) and 9-rhombohedral (9R)-polytype, when compared to the twin boundaries and the intrinsic stacking fault (iSFs) [35]. Some single twin planes and arrays of {111} stacking faults were

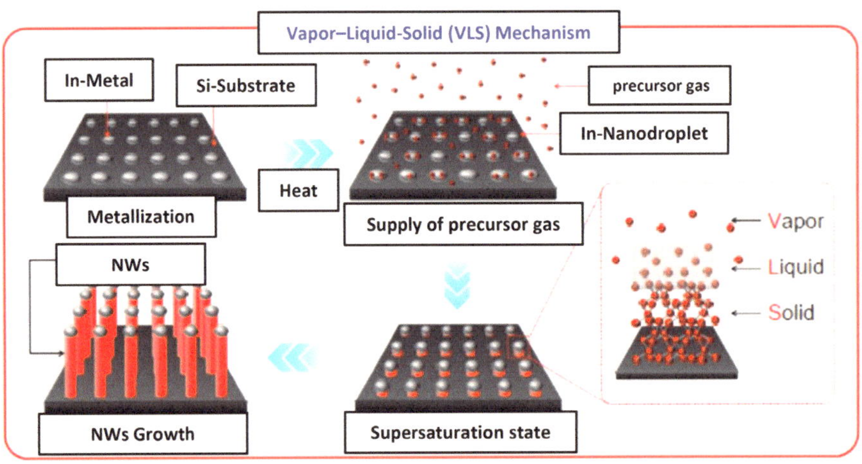

Figure 1.
Schematic flow mechanisms of In-catalyzed Si NWs grown by VLS growth mode.

observed in Au-catalyzed Si NWs grown in the ⟨112⟩ direction [14]. Lopez *et al.* reported about such structures in Au-catalyzed ⟨111⟩-oriented kinked shaped Si NWs [36]. Such defects were found to be running parallel to the ⟨112⟩ NW axis, and often extending in the entire length of the wire. There is no report about the stacking fault as well as planar defects (twin planar defects) along ⟨112⟩ direction in the case of In-catalyzed, vertically aligned, and (111)-oriented Si NWs grown by VLS mode. Our main objective is to control the verticality of p-type Si NWs for the applications of nano-devices. We also investigate about the single Si NW, whether it contains twining defects, planar defects as well as stacking fault along the growth direction of In-catalyzed Si NWs. Second, we attempted to present one simple model about the root cause of stacking faults formation that are distributed around the Si NWs as well as at the interface of In-NDs/Si-substrate (polytypes). Finally, it was established that the Si NWs height were also restricted by the In-NPs migration from the top of the Si NWs.

2. Experimental methods

The basic mechanism behind the VLS growth mode is the transformation of the solid metal catalyst nanoparticle into a liquid alloy of the catalyst and compound of the semiconductor. In this case the liquid particle acts as a privileged site for Si deposition (precipitation via liquid catalyst), and has higher sticking coefficient as compared to the solid surfaces, shown in the schematic flow mechanism of **Figure 1** [37]. Two different type of experiments were conducted to grow Si NWs. First, before the air-breaking condition p-type 300 μm-thick Cz-Si (111) substrate having resistivity of 1–10 Ω-cm, was cleaned by RCA washing. Next, In-NDs were grown on Si (111) substrate, using a conventional thermal evaporation system by evaporating pure In wire with base pressure (P_B) of 4.4×10^{-4} Pa. Stranski–Krastanow (SK) growth mode was followed by In-NDs, which is not a two-dimensional growth, but rather gives rise to islands of the In-metal. Subsequently, the In-NDs were thermally annealed in a glass tube furnace at 630°C for 360 min and then treated by a H_2 plasma at substrate temperature (T_S) of 200°C for 30 min under a pressure of 10 Pa in a sputtering chamber (after air-breaking condition). Finally, Si NWs were grown by a radio frequency (RF) magnetron sputtering after air-breaking condition (sample-Na) at T_S = 630°C under pressure of 1 Pa for 30 min growth time.

In the next experiment everything was grown in the same sputtering chamber (without air-breaking condition). First, 300 μm-thick p-type Cz-Si (111) substrate, having resistivity ~1–10 Ω-cm was washed by RCA washing. Soon after RCA washing and drying the wafer was transferred to the plasma assisted and high vacuum sputtering chamber having background pressure, $P_B \sim 6.0 \times 10^{-6}$ Pa. Secondly, the In-atoms were deposited on Si substrate by In-sputtering target at room temperature (RT) under working pressure of 3 Pa for 20 min. Thirdly, the as-grown In-Islands sample was treated by H_2-plasma (200 sccm) in Ar (20 sccm) environment, where substrate temperature was set to 600°C, for 3 min. The heating rate of the substrate was kept 10°C/min to get self-organized and well defined In-NDs on the substrate before the Si NWs growth. Finally, the substrate temperature was set to 600°C, assisted by H_2-plasma (200 sccm) in Ar (20 sccm) for 60 min under working pressure of 10 Pa and Si NWs were successfully grown (sample-Nw).

The interface scenario between In-NDs and Si (111), surface morphologies, shape, density, and contact angle (θ_C) of In-NDs on Si-substrate were observed by high resolution scanning electron microscopy (HR-SEM). Crystal structure, i.e., cross-section, different planar defects and twining of planar defects in the entire Si NWs were observed, via selected-area of electron diffraction (SAED) as well as via cross-sectional view by high-resolution transmission electron microscope (HR-TEM, Model: H-9000NAR). Migration and trapping of In-Nanoparticles (NPs) on the side wall of Si NWs were deeply investigated by energy-dispersive X-ray spectroscopy (EDX) and dark field scanning of scanning-TEM (STEM).

3. Results and discussion

Indium metal has a low melting point, and the eutectic temperature of the In–Si binary system coincides with the melting temperature of indium at 157°C [20, 38]. It is also worthy to note that the In–Si eutectic alloy exhibits a steep liquidus line, such that the liquid alloy can promptly be supersaturated with Si in a wide range of temperatures (at least up to 800°C) and In-Si eutectic alloy has an extremely low Si solubility approximately ~10^{-4} at.% Si [26].

In this work the In-NDs were grown with the optimized growth condition, where the In-deposition time were increased from 20 min to 30 min in the plasma assisted crystal growth reactor as shown in **Figure 2a** and **b**. By decreasing the T_S from 600 to 530°C in the high vacuum along with H_2-plasma treatment, the diameter of the In-NDs were varied from 45 nm and to 500 nm. However, in some In-NDs, got an exceptionally low contact angle as shown in **Figure 2b**, which is quite interesting in the context of NWs verticality control for better Si-precipitation via In-NDs. On the other hand, we observed exceedingly high contact angle approximately 140°, as shown in **Figure 2a**. In this case the liquid NDs acts as a facilitator site for Si deposition (precipitation). It was found that In-NDs have higher sticking coefficient than the solid surfaces [37]. As shown in the schematic view of **Figures 1** and **2c**, where the supersaturation of the In-NDs, induced by the continuous gas phase supply of Si species (Vapor), leads to the precipitation of Si nanowires (Solid) at the interface of Si-substrate and In-NDs (Liquid). Si NWs growth was initiated, when a steady-state condition between the flux of the Si through the particle and the precipitation of Si on the substrate via In-NDs was reached, shown in **Figure 1** [26, 37]. Later the growth condition for Si NWs was improved from sample-Na to sample-Nw. The contact angle along with the In-NDs size together can define the diameter of the Si NWs as depicted in the **Figures 2c** and **3a, b**. It is all about the growth condition as well as growth in the reactor with and without air-breaking. By using HR-SEM, we investigated the interface condition between the In-NDs and the Si substrate prior to the crystal growth of Si NWs, shown in **Figure 3a** and **b**. In the case of thermally evaporated In-NDs on Si-substrate, a spherical shaped In-NDs were observed with different sizes in the range of 30–100 nm, as well as with quite large θ_C of 140°, shown in **Figure 3a**.

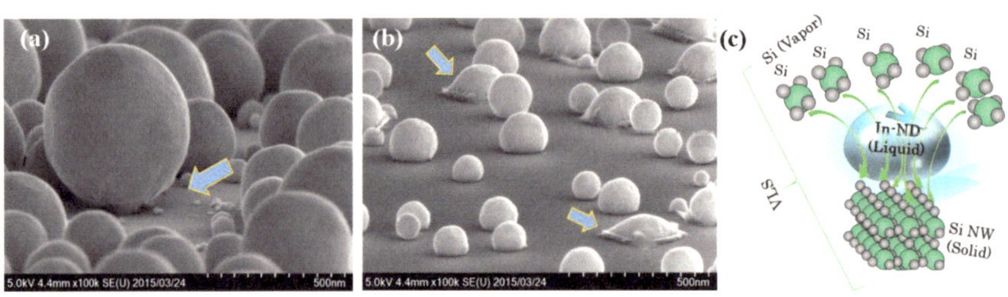

Figure 2.
(a) SEM images of In-NDs (high contact angle) grown by sputtering of In-target at RT, for 20 min and then treated by plasma treatment at 530°C for 3 min, (b) SEM images of In-NDs (low contact angle) grown by sputtering of In-target at RT, for 20 min and then treated by H_2-plasma at 600°C for 3 min, and (c) Si precipitation mechanism in the VLS growth mode via well wetted In-NDs to grow Si NWs.

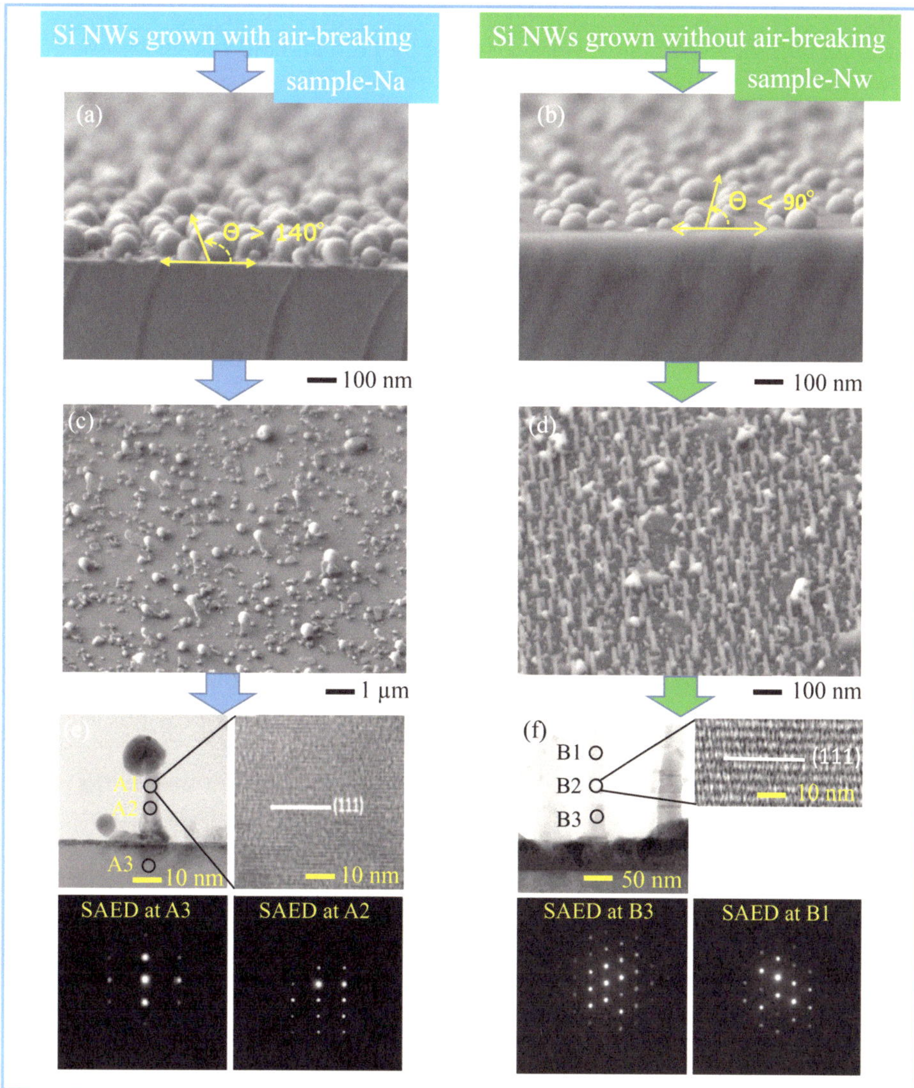

Figure 3.
SEM images of In-NDs grown by sputtering in the reactor, (a) with air-breaking (sample-Na), (b) without air-breaking (sample-Nw). TEM micrograph of Si NWs grown in the reactor, (c) with air-breaking, (d) without air-breaking. HR-TEM micrograph of the as-grown Si NWs (e) taken at point "A1" in the sample-Na, and the SAED pattern taken at point "A3" and "A2" are shown in the inset, and (f) taken at point "B2" in the sample-Nw, and the SAED pattern taken at point "B3" and "B1" are shown in the inset. ***Figure 3*** *reproduced with the permission from ref [39]. Copyright 2015 the Royal Society of Chemistry (RSC).*

The spherical shape of In-NDs (due to low wettability) on the Si substrate can be explained by the interactions mechanism of oxidized In-NDs (In_2O_3) in the context of surface free energy (E_f) and θ_C in a qualitative manner on Si substrate in the conventional physical evaporator. Previously, the surface free energy (E_f) for pure In droplet and In_2O_3, respectively, were found to be 525 mN/m and 500–520 mN/m at 850 K [40]. During the growth of In-NDs in the air-breaking scenario, where the E_f of In-NDs on Si might be reduced due to the presence of the thin oxide layer around the In-NDs on substrate and subsequently increased to the θ_C up to ~140° of In-NDs on Si substrate, shown in **Figure 3a**. As a result, extremely low density of vertically aligned (111)-Si NWs at the supersaturation phase were grown, shown in **Figure 3c**. In the grown Si NWs, one can see quite big used cap of In-NDs on the top of Si NWs. The non-uniform diameter and length of Si NWs ranging for 30–100 nm and 150–200 nm, respectively, were observed by TEM and SEM observations, shown in **Figure 3c**. It can be speculated that the precipitation of Si atomic migration through the liquid phase of In-NDs to the Si substrate at the eutectic phase might be hindered by the thin oxide layer that exists around the majority of In-NDs and on the Si substrate. This kind of low wettability condition with extremely large θ_C of In-NDs on Si substrate does not support to the crystal growth of high-density Si NWs, shown in **Figure 3c**. Due to the small contact area of wet surface of In-NDs on the Si substrate may not be symmetric too, which ultimately give rise to the tilted Si NWs with big In-NPs cap, shown in **Figure 3c**. The (111)-oriented crystallinity at point "A1" was confirmed by the HR-TEM image (**Figure 3e**). The SAED at points "A3" and "A2" (shown in the inset of **Figure 3e**) corresponding to Si NWs and Si substrate, respectively, exhibit a spotty pattern, which indicating that the VLS growth was happened successfully for few Si NWs (with extremely low density of 2.5 μm^{-2}) using crystal growth in the air-breaking condition (sample-Na).

In the second experiment (without air-breaking), we optimized the growth conditions to obtain good interface between In-NDs and Si substrate by eliminating the oxide layer to enhance the wettability. All the in-situ growth steps starting from In-NDs growth, plasma treatment, and up to the crystal growth of Si NWs were performed in a relatively clean and high vacuum chamber without air-breaking (sample-Nw). The θ_C of In on Si (111) was reported to be approximately 125° at 350°C by Mattila *et al.* [41] After improving the wettability condition, the θ_C of In-NDs on Si substrate was remarkably decreased from our previous value of 140° to 80°, shown in the inset of **Figure 3b**. Ultimately the In-NDs on the Si substrate were suppressed to hemispherical shaped with uniform sizes in the range of 70–100 nm (without air-breaking). The wettability of the In-NDs and increase of the E_f of In-NDs, are attributed to the combined action of the H_2-plasma treatment at T_S of 600°C, and clean growth environment by performing the entire growth process in the same chamber (without air-breaking). The verticality of (111)-oriented Si NWs was nicely controlled with a uniform top diameter and length, respectively, of approximately 18 nm and 100 nm, shown in **Figure 3d** and **f**. The top diameter value approximately ~18 nm of Si NWs is the smallest value ever achieved for the applications of nano-scale devices application. The controllability of the verticality of Si NWs are attributed to the cylindrically symmetric flow of the precipitated Si atoms via highly wettable In-NDs toward the Si substrate contact points at an incredibly low θ_C, shown in **Figure 3d–f**. The density of the grown Si NWs at T_S = 600°C was also remarkably enhanced by 28 time from 2.5 μm^{-2} (sample-Na) to 70 μm^{-2} (sample-Nw). The Si NWs were found to be cone shaped after tapering, which is caused by the migration of In-NPs from the top of NWs, shown in **Figure 3a** and **b**. Based on the HR-TEM observation at point "B2", the (111)-orientation of Si NWs was confirmed (as shown in the inset of **Figure 3f**). Also, spotty pattern both in Si NWs and Si substrate at point "B1" and "B3", respectively, were confirmed by SAED observation (shown in the inset of **Figure 3f**). Energy dispersive X-ray spectroscopy (EDX) was taken near the cap of the Si NWs to investigate the compositional information. It was found that NW only contains a Si peak, while In was detected for the spheres only. The top diameter value approximately 18 nm was found quite small (without air-breaking) when compared to the previously obtained vertically aligned Si NWs grown with In catalyst in the VLS mode (with air-breaking condition). We strongly need to decrease the tapering phenomena and also to increase the length of the In-catalyzed Si NWs. These are still an open research challenges and it can be speculated that the control of the diameter of In-NDs both at the top and base area might be possible after suppressing the trapping rate of In NPs by Si NWs. Further reduction of In-NDs θ_C on Si substrate can further enhance the length of NWs. We need to further reduce the size and wettability of In-NDs and then can further reduce the diameter of the vertically aligned Si NWs to realize the quantum size effect for wide band gap applications of solar cells. In-catalyzed Si NWs may be a potential candidate material for nanoscale devices, because of its shallow acceptor levels and its vertical alignment.

Alet *et al.* grown In-catalyzed crystalline Si NWs at low temperature on Indium doped Tin Oxide (ITO), which were not vertically aligned and detail investigation about the planar as well as stacking fault were not given [42]. In our case a typical cross-section TEM image of the cone like Si NWs arrays grown at 600°C has been shown in **Figure 4b**. The HR-TEM shown in **Figures 3d** and **4b** shows that the Si NWs have an average diameter of ≅ 18 nm at the top and average diameter≅ 30 nm at the bottom, lengths longer than 100 nm and grow at a right angle with respect

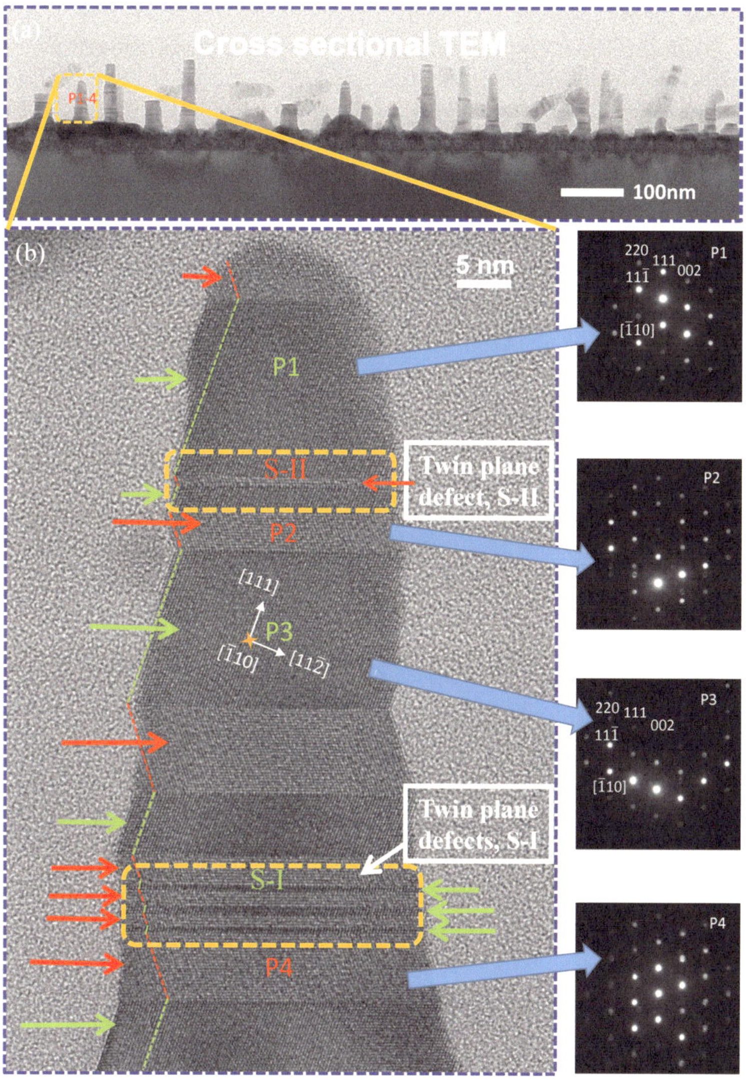

Figure 4.
(a) Cross-sectional TEM image of the grown Si NWs on Si (111), (b) HR-cross-sectional TEM image of an individual Si NW from fig. 4a (In the inset of (b), the corresponding SAED pattern taken along the [110] zone axis at point “P1”, “P2”, “P3”, and “P4” respectively are given). Different planar defects along the axial segments and twining of planar defects has been shown at point S-I and S-II, respectively. ***Figure 4*** *reproduced with the permission from ref [43]. Copyright 2016 the Japan Society of Applied Physics (JSAP).*

to the Si (111) surface as shown in **Figures 3d–f** and **4b**. The shape of the Si NWs were found to be cone like, and this can be explained by the similar analogy of tapering of the NWs grown by Au-catalyzed [44, 45]. Sharma *et al*. found that, when NWs elongate from the In-NDs (or In- droplets), the base area of the NWs remains exposed for a longer time to the reactive radicals when compared to the newly grown upper part of the NWs [46]. In the case of Au-catalyzed Si NWs grown in the ⟨112⟩ direction, where single twin planes and arrays of {111} stacking faults were reported [14], whereas Lopez *et al.* found the same structures for ⟨111⟩-oriented Au-catalyzed grown Si NWs [36]. These defects orientation is parallel to the ⟨112⟩ NW axis, often extending throughout the entire length of the wire. There is no report about the crystalline twining defects, planar defects and stacking fault along ⟨112⟩ direction in case of In-catalyzed vertically aligned, and (111)-oriented Si NWs.

We can see the planar defects as well as twining defects appearing in many segments of the Si NWs (sample-Nw) as shown in the inset of **Figure 5b** by green color arrow as well as red color arrow along the ⟨112⟩ direction. The (111)-oriented Si NWs at segments "P1–4" were also confirmed by HR-TEM image, as shown in **Figure 4b**. The corresponding selected area diffraction (SAED) shown in the in-set of **Figure 4b** has been recorded along the [110] zone axis from the Si NWs and which confirm the single crystal nature and its axial direction. But you can see twining of planes along ⟨112⟩ direction, which are distributed along the axial direction marked as S-I and S-II, as shown in **Figure 4b** for single (111)-oriented Si NW.

It can be established that the HR-TEM image of single Si NWs along with the SAED at "P1" and "P3" of Si NWs (taken from **Figure 5b**), marked by green color

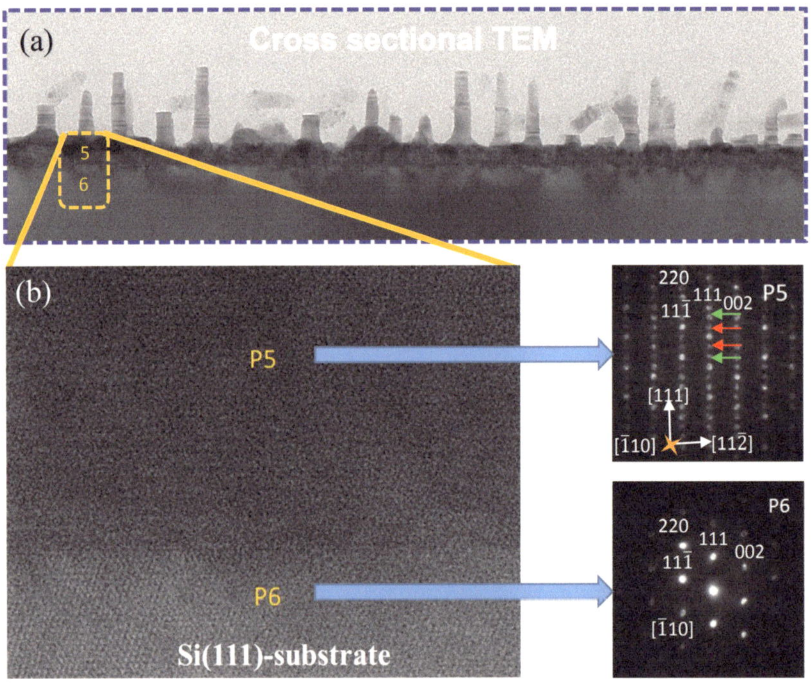

Figure 5.
(a) High-resolution cross-sectional TEM image around the Si-substrate and deposited In-NDs surrounded by Si NWs at the interface, (b) the corresponding SAED pattern taken along the [1 $\bar{1}$ 0] zone axis at point "P5" and "P6" respectively. In the inset of (b) is a SAED along [$\bar{1}$ 10] zone axis at point "P5", where the red arrow intensity originates from 1/2{111} spots associated with the 2H polytype and the green intensity is originating from 1/3{111} spots associated with 9R stacking. Figure 5 reproduced with the permission from ref [43]. Copyright 2016 the Japan Society of Applied Physics (JSAP).

arrow, follow the same crystallinity of Si-substrate orientation marked at "P6" in **Figure 5b**. The diamond structure of the Si NWs with a lattice constant = 0.543 nm was confirmed by HR-TEM image and the SAED pattern at segments "P1" and "P3", which corresponds to the space group Fd3m grown along the ⟨111⟩ direction. Unfortunately, the SAED at "P2" and "P4" (taken from **Figure 4b**) of Si NW shows small tilt with respect to orientation of Si-substrate, and the same have been marked by red color arrow, as shown in HR-TEM image given in **Figure 5b**. Spotty pattern both at "P2" and "P4" segments of the NW were confirmed by SAED observation, as shown in the inset of **Figure 5b** and same value of small angular tilt of both "P2" and "P4" segments were observed. We concluded that the Si NWs are vertically aligned except the twining of planar defects, which might be caused by the faster cooling rate 100°C/6 min, as well as due to the longer exposer time of downside wall to the reactive radicals as compared to the upper side of Si NWs [46]. We have to consider about the thermal conductivity mechanism, which is different for Si NWs as compared to the bulk Si-wafer [35].

Here we focus to discuss about the Si NWs crystallinity at the interface of emanating Si NWs from the Si-substrate, as shown in **Figure 5a**. **Figure 5b** gives the SAED pattern at point "P5", which confirm the stacking fault in the grown Si NWs. The red arrow intensity originates from 1/2{111} spots is related to the 2H-polytype (stacking ABAB...) and the green arrow intensity originates from 1/3{111} spots is related to 9R (stacking ABCBCACAB...) [47].

Such structures are either attributed to scattering phenomena from two overlapping crystals with a stepped {111} twin boundary (parallel to the electron beam) [48, 49], or it might be the direct evidence of a 9R-polytype [23]. Furthermore, it has been established, when the crystallographic direction of the lattice abruptly changes in the In-Si material system then stacking fault may generated. Especially, when two crystals parts begins to grow separately and then meet at certain point, where the crystallographic direction remains the same, but each side of the boundary has an opposite phase. These kinds of stacking fault can be formed due to the complex dynamics of the In-NPs migration as well as mixing of the Si-atoms from the top of the Si NWs as well as precipitated Si atoms too. Such complex scenario will be explained later with the help of model given in **Figure 7**.

We realized that the In-NPs migration from the top of Si NWs toward the unused In-NDs and Si-substrate interface, where Si NWs are emanating from Si-substrate may cause to the stacking fault. We also know that an isolated defect like {111} faults have been observed to trap Au-atoms [34]. Therefore one cannot negate the possibility of In-atoms trapping by the planar defects of Si NWs during the VLS mode growth. **Figure 6a** gives the dark field-STEM image of the Si NWs, where one can clearly see the white spherical contrast of In-NPs around the side wall of the Si NWs. The compositional investigation by EDX taken at top of the Si NWs as well as taken from the pure In-NPs around the side wall of the Si NWs has been shown in the **Figure 6b**. The Kα X-ray energies for the In is 24.21 keV, and Lα X-ray energies for the same elements is 3.287 keV. Lα lines of the In-atoms can be separated, and this technique can be quantitatively used in a SEM. As shown in **Figure 6c** and **d**, we observed a reasonable mapping of Si and In sources, which originate from the Si NWs top and from the pure spherical In-NPs on the side wall of the Si NWs structure, respectively. The same phenomenon for Au NPs on the side of the Si NWs was observed by Krylyuk *et al.* [44]. Migration of In-NDs from the top of NWs was observed and confirmed by both dark field and bright field-STEM images. The coalescence of the migrated In-NPs with the unused In-NDs having high contact angle may resulted in relatively large size (~ 500 nm) of In-island as shown in the red color dotted marked line in **Figure 6a** and **d**, respectively. The generation mechanism will be explained in the proposed model given in **Figure 7**.

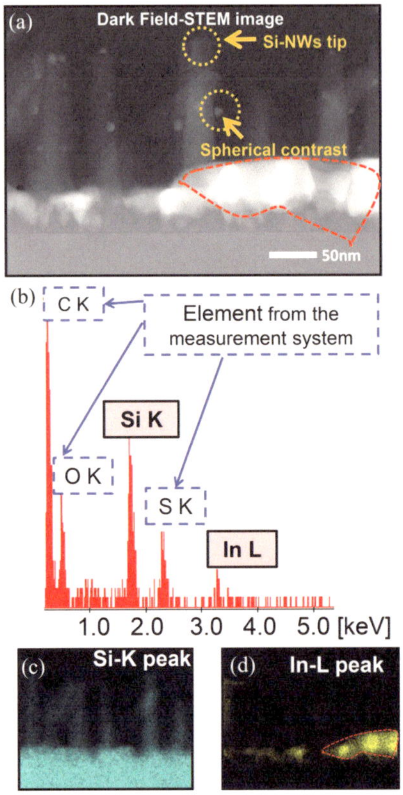

Figure 6.
(a) Dark-field-STEM images of Si NWs along the [1 $\bar{1}$ 0] zone axis to show the In-NPs as a white spherical contrast on the side wall of the Si NWs, (b) Elemental analysis of sample-Nw, (c) EDX intensity mapping of pure Si NWs, and (d) EDX intensity mapping of precipitated In-composition. Figure: 6 reproduced with the permission from ref [43]. Copyright 2016 The Japan Society of Applied Physics (JSAP).

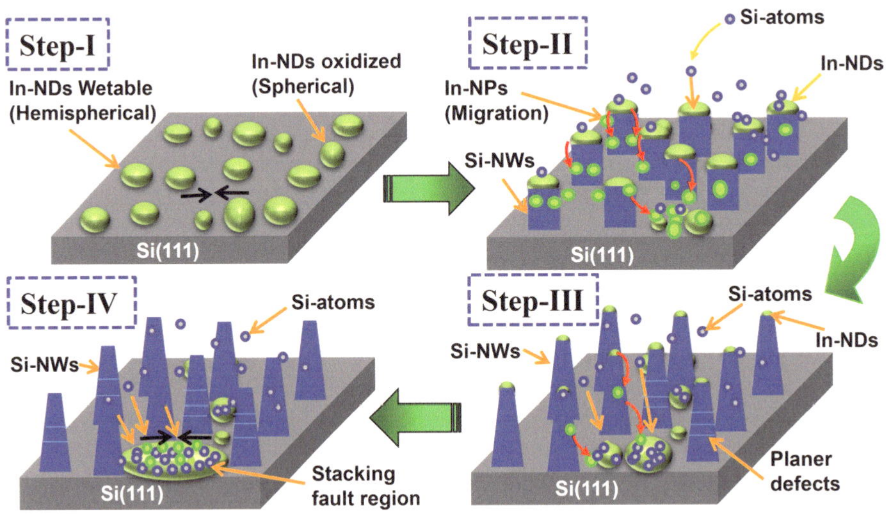

Figure 7.
Schematic overview of the proposed model to explain step by step about the In-NPs migration from the top of the Si NWs and tapering as well as stacking fault formation in Si NWs, which in return restricted the length of the Si NWs. Figure 7 is reproduced with the permission from ref [43]. Copyright 2016 the Japan Society of Applied Physics (JSAP).

The reason of the In-NPs migration from the top of the Si NWs toward the Si-substrate during the crystal growth of Si NWs may not be clearly known yet but we can anticipate the mechanism of In-NPs coalescence with In-NDs and mixing with precipitated Si-atoms via proposed model shown in **Figure** 7.

The In-catalyzed Si NWs grown by VLS mechanism confronted with planar defects, twining and stacking fault, which were observed by HR-STEM, as shown in **Figures 4a, b** and **5a, b**. To explain about the In-NPs migration to Si-substrate, Si NWs tapering, stacking fault, as well as limiting of NWs length by In-NPs migration, one simple model was anticipated. In this model, as a first step (Step-I), the In-NPs on Si(111), investigated, where most of the In-NDs got good wettability after Ar/H_2-plasma treatment at 600°C with semispherical shape and few In-NDs were got spherical shape as shown in **Figure 2a** as well as in **Figure 3a** and the same is depicted schematically in **Figure** 7, Step-I. The spherical shape In-NDs may be confronted with thin oxide problems at the interface, shown in **Figure** 7 (Step-I). During the Si NWs growth as a second step (Step-II) we observed that the In-NPs migration from the top of Si NWs toward the Si-substrate are taking place and mixing with already grown In-NDs having thin oxide interface layer between Si-substrate and In-NDs, as shown in **Figure 6a–d** and the same is depicted schematically in **Figure** 7. Due to the migration of In-NPs toward the Si-substrate some new contact points on Si-substrate were created and the already In-NDs size were enhanced or elongated.

The supersaturation of the droplet, induced by the continuous gas phase supply of Si species, leads to the precipitation of Si nanowhiskers at the interface between the particle and the substrate, shown in **Figure 1**. Growth is obtained, as shown in **Figure** 7, Step-II, when a steady-state condition is reached between the flux of the Si through the In-NDs and the precipitation of Si on the Si-substrate [26, 37]. At the same time Si-atoms from sputtering source are also adsorbed on the same surface of In-NDs already deposited In-NDs on substrate having weak oxide interface layer, where the super saturation limit may be exceeded and now the precipitation of In-NPs were initiated as depicted in **Figure** 7, Step-II and the same has been confirmed in **Figure 6a–d**. Previously, the boron precipitation limit in $BaSi_2$ were studied by in-plane and out of plane XRD, HR-STEM and TEM measurement and then successfully overcome the boron precipitation issue in bulk thin film *p*-$BaSi_2$ layer [50, 51]. Altogether XRD out of-plane and in-plane characterization, SEM and HR-STEM observation could be a useful approach to investigate the precipitation of In-NPs in Si NWs. In Step-III, tapering of NWs are started due to the decreasing of In-NDs size at the top of Si NWs, which is caused by the In-NPs migration/trapping from the top of the Si NWs as well as due to the longer exposer time of downside wall to the reactive radicals as compared to the upper side of Si NWs, as shown in **Figure 6a** and **c** [46]. At the same time the Si atoms adsorption by In-NDs are increasing because the In-NDs at the top of the Si NWs are decreasing. During such situation, the adsorbed Si-atoms by In-NPs cannot be precipitated to grow Si NWs due the partial oxide thin layer between the In-NDs and Si-substrate but instead support to sidewise expansion of the two neighboring In-NDs, shown in **Figure 6a–d** and the same has been depicted in Step-III, of **Figure** 7. At last, the two In-NDs coalesce on the surface of Si-substrate surrounded by Si NWs, as depicted in schematically in Step-III of **Figure** 7. Also, the deposition of the Si-atoms is increasing and at the same time the In-NPs migration are decreasing, which initiated the stacking fault shown in the inset of **Figures 4a, b** and **5a, b**. In Step-IV, finally the In-NPs migration is stopped as shown in **Figure 6a** as well as depicted in **Figure** 7. Subsequently, highly precipitated of In-NPs/NDs in the liquid phase are mixed with Si-atoms then finally give rise to stacking fault at the Si-substrate surrounded by

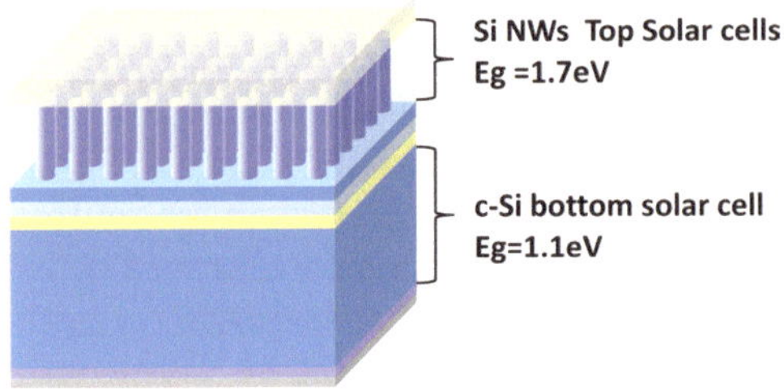

Figure 8.
Schematic Si NWs/c-Si tandem solar cells with expected efficiency of 30% (1sun).

the neighboring Si NWs and the same has been confirmed in the inset of **Figure 5b** at "P5". Probably at this situation the sticking coefficient of Si-atoms are reduced, and no further Si-atoms can be absorbed by In-NDs island. This is the time where all In-NDs at the top of Si NWs might be disappeared and give rise to stacking fault on the surface of Si-substrate surrounded by Si NWs, shown in **Figure 6a** as a white contrast (red dotted line). As a result, Si NWs growth were stopped, and we can safely say that the tapering of the Si NWs and length of the Si NWs were restricted by In-NDs disappearance or migration from the top of the In-catalyzed Si NWs.

Suppression of the In-NPs migration from the top of the Si NWs are essential to grow longer Si NWs as well as to avoid NWs tapering and also to fix the stacking fault in Si NWs. We need to find new growth condition to suppress the In-NPs migration to find the lower optimal substrate temperature without compromising on the verticality control of Si NWs. The ultra-clean interface between Si-substrate and In-NDs is essential to get smaller θ_C of the In-NDs to increase its wettability on Si-substrate, without any oxidation issues prior to Si NWs growth. We have to try with longer Ar/H_2 plasma treatment time, by using low plasma power to avoid any kind of surface damage to the Si-substrate to further reduce the θ_C. Our new growth condition can be used to grow vertically aligned Si NWs using In-catalyzed in VLS mode for many electronic and nano device applications [52]. Solar cell architecture of the 4-terminal based wide bandgap top cell (Si NWs) and narrow bandgap bottom cell for best matching efficiency with 10% Ge in SiGe active layer is possible, where the photocurrent limits under the solar spectrum for varying band gap of SiGe materials due to the Ge composition for bottom Heterojunction solar cell applications can be achieved [53]. Heterojunction light-emitting diodes (LEDs) comprising p-type Si nanowires (p-Si NWs) and n-type indium gallium zinc oxide (n-IGZO) were successfully fabricated [54]. Band gap energy of Si NWs can be controlled around 1.7 eV by changing the diameter of the NW [3]. Quite high conversion efficiency around 30% is expected in the Si NWs/c-Si tandem solar cells structure, as shown in **Figure 8**.

4. Summary

Using In-catalyzed VLS mode growth, we have successfully controlled the verticality and (111)-orientation of Si NWs and ultimately scaled down the diameter of NWs to 18 nm. The density of vertically aligned Si NWs was enhanced from 2.5 μm^{-2} to 70 μm^{-2}. During the *in situ* sequential deposition of In-NDs catalyzed Si NWs in high vacuum environment has successfully blessed us with vertically aligned and (111)-oriented Si NWs arrays using VLS mode. Using the HR-TEM, HR-SEM, and EDX, the planar defects as well as twining defect structure, which is grown perpendicularly to the Si-substrate (along ⟨112⟩ Si-NW direction) to the axial direction (111)-orientation in many parts of the individual Si NWs were found. Stacking fault were confirmed, where the intensity originates from 1/2{111} spots associated with the 2H polytype and also the intensity originates from 1/3{111} spots associated with 9R stacking in the emanating NW crystal structure at the interface of Si-substrate. These kinds of studies could greatly influence the future understanding about the phase purity, crystallinity and controlled growth of In-catalyzed Si NWs, especially the length and periods of vertically aligned Si NWs for the Nano device applications including solar cells and LEDs.

Acknowledgements

This work was partially supported by the MEXT, FUTURE-PV Innovation (FUkushima Top-level United Center for Renewable Energy Research–Photovoltaics Innovation) Project.

Author details

M. Ajmal Khan[1]* and Yasuaki Ishikawa[2]

1 RIKEN Cluster for Pioneering Research (CPR), Wako, Saitama, Japan

2 Department of Electrical Engineering and Electronics, College of Science and Engineering, Aoyama Gakuin University, Sagamihara-shi, Kanagawa, Japan

*Address all correspondence to: muhammad.khan@riken.jp; m_ajmal_khan@yahoo.com

References

[1] R. P. Feynman: Engineering and Science **23** (1960) 22.

[2] Y. Kurokawa, S. Kato, Y. Watanabe, A. Yamada, M. Konagai, Y. Ohta, Y. Niwa, and M. Horita: Jpn. J. Appl. Phys. **51** (2012) 11PE12.

[3] D. D. D. Ma, C. S. Lee, F. C. K. Au, S. Y. Tong, and S. T. Lee: Science **299** (2003) 1874.

[4] B. Tian, X. Zheng, T. Kempa, Y. Fang, N. Yu, G. Yu, J. Huang, and C. M. Lieber: Nature **449** (2007) 885.

[5] L. Tsakalakos, J. Balch, J. Fronheiser, B. A. Korevaar, O. Sulima, and J. Rand: Appl. Phys. Lett. **91** (2007) 233117.

[6] T. Ernst: Science **340** (2013) 1414.

[7] Z. Li, Y. Chen, X. Li, T. I. Kamins, K. Nauka, and R. S. Williams: Nano Lett. **4**, (2007) 245.

[8] Y. Cui, Q. Wei, H. Park, and C. M. Lieber: Science **293** (2001) 1289.

[9] A. I. Hochbaum, R. Chen, R. D. Delgado, W. Liang, E. C. Garnett, M. Najarian, A. Majumdar, and P. Yang: Nature **451** (2008) 163.

[10] R. E. Algra, M. A. Verheijen, M. T. Borgström, L.-F. Feiner, G. Immink, W. J. P. van Enckevort, E. Vlieg, and E. P. A. M. Bakkers: Nature **456** (2008) 369.

[11] P. Caroff, K. A. Dick, J. Johansson, M. E. Messing, K. Deppert, and L. Samuelson: Nat. Nanotechnol. **4** (2009) 50.

[12] A. Fissel, E. Bugiel, C. R. Wang, and H. J. Osten: Mater. Sci. Eng: B **134** (2006) 138.

[13] J. Arbiol, A. Fontcuberta i Morral, S. Estradé, F. Peiró, B. Kalache, P. Roca i Cabarrocas, and J. R. Morante: J. Appl. Phys. **104** (2008) 064312.

[14] F. M. Davidson, III, D. C. Lee, D. D. Fanfair, and B. A. Korgel: J. Phys. Chem. C **111** (2007) 2929.

[15] Z. W. Wang and Z. Y. Li: Nano Lett. **9** (2009) 1467.

[16] S. K. Srivastava, D. Kumar, P. K. Singh, and V. Kumar: Proc. 34th IEEE Photovoltaic Specialists Conference, 2009, p. 1851.

[17] R. S. Wagner and W. C. Ellis: Appl. Phys. Lett. **4** (1964) 89.

[18] S. Iijima: Nature **354** (1991) 56.

[19] M. S. Jeon and K. Kamisako: Mater. Lett. **62** (2008) 3903.

[20] R. W. Olesinski and G. J. Abbaschian: Bulletin of Alloy Phase Diagrams **5** (1984) 273.

[21] E. I. Givargizov: J. Cryst. Growth **31** (1975) 20.

[22] Y. Wang, V. Schmidt, S. Senz, and U. Gösele: Nat. Nanotechnol. **1** (2006) 186.

[23] M. K. Sunkara, S. Sharma, R. Miranda, G. Lian, and E. C. Dickey: Appl. Phys. Lett. **79** (2001) 1546.

[24] J. Zhang, F. Jiang, Y. Yang, and J. Li: J. Cryst. Growth **307** (2007) 76.

[25] J.-Y. Yu, S.-W. Chung, and J. R. Heath: J. Phys. Chem. B **104** (2000) 11864.

[26] F. Iacopi, P. M. Vereecken, M. Schaeker, M. Caymax, N. Moelans, B. Blanpain, O. Richard, C. Detavemier, and H. Griffiths: Nanotechnol. **18** (2007) 505307.

[27] M. Jeon and K. Kamisako: J. Nanosci. Nanotechnol. **8** (2008) 5188.

[28] M. Jeon, Y. Tomitsuka, and K. Kamisako: J. Ind. Eng. Chem. **14** (2008) 836.

[29] M. Jeon and K. Kamisako: J. Alloys Compd. **476** (2009) 84.

[30] T. David, D. Buttard T. Schulli, F. Dalhüin, and P. Gentile: Surf. Sci. **602** (2008) 2675.

[31] F. Li, P. D. Nellist, and D. J. H. Cockayne: Appl. Phys. Lett. **94** (2009) 263111.

[32] Y. Ohno, N. Yamamoto, K. Shoda, and S. Takeda: Jpn. J. Appl. Phys. **46** (2007) L830.

[33] H. Hibino, K. Sumitomo, and T. Ogino: J. Vac. Sci. Technol. A **16** (1998) 1934.

[34] J. E. Allen, E. R. Hemesath, D. E. Perea, J. L. Lensch-Falk, Z. Y. Li, F. Yin, M. H. Gass, P. Wang, A. L. Bleloch, R. E. Palmer, and L. J. Lauhon: Nat. Nanotechnol. **3** (2008) 168.

[35] H. F. Zhan, Y. Y. Zhang, J. M. Bell, and Y. T. Gu: J. Phys. D: Appl. Phys. **47** (2014) 015303.

[36] F. J. Lopez, E. R. Hemesath, and L. J. Lauhon: Nano Lett. **9** (2009) 2774.

[37] A. P. Levitt: 1970 *Whisker Technology* (Wiley, New York, 1970) pp. 47-119.

[38] T. B. Massalski: Binary Alloy Phase Diagrams (Metals Park, Ohio, American Society for Metals, 1990) 2nd ed., pp. 505-763.

[39] M. Ajmal Khan, Y. Ishikawa, I. Kita, K. Fukunaga, T. Fuyuki, and M. Konagai: J. Mater. Chem. C **3** (2015) 11577.

[40] E. Ricci, T. Lanata, D. Giuranno, and E. Arato: J. Mater. Sci. **43** (2008) 2971.

[41] M. Mattila, T. Hakkarainen, H. Lipsanen, H. Jiang, and E. I. Kauppinen: Appl. Phys. Lett. **89** (2006) 063119.

[42] P.-J. Alet, L. Yu, G. Patriarche, S. Palacin, and P. Roca i Cabarrocas: J. Mater. Chem. **18** (2008) 5187.

[43] M. Ajmal Khan, Y. Ishikawa, I. Kita, A. Tani, H. Yano, T. Fuyuki, and M. Konagai: Jpn. J. Appl. Phys. **55** (2016) 01AE03.

[44] S. Krylyuk, A. V. Davydov, and L. Levin: ACS Nano **5** (2011) 656.

[45] J. B. Hannon, S. Kodambaka, F. M. Ross, and R. M. Tromp: Nature **440** (2006) 69.

[46] S. Sharma and M. K. Sunkara: Nanotechnol. **15** (2004) 130.

[47] H. Cerva: J. Mater. Res. **6** (1991) 2324.

[48] C. Cayron, M. D. Hertog, L. Latu-Romain, C. Mouchet, C. Secouard, J.-L. Rouviere, E. Rouviere, and J.-P. Simonato: J. Appl. Cryst. **42** (2009) 242.

[49] H. Bender, A. De Veirman, J. Van Landuyt, and S. Amelinckx: Appl. Phys. A: Mater. Sci. Process **39** (1986) 83.

[50] M. Ajmal Khan, K. O. Hara, K. Nakamura, W. Du, M. Baba, K. Toh, M. Suzuno, K. Toko, N. Usami, and T. Suemasu: J. Cryst. Growth **378** (2013) 201.

[51] M. Ajmal Khan, K. Nakamura, W. Du, K. Toko, N. Usami, and T. Suemasu: Appl. Phys. Lett. **104** (2014) 252104.

[52] M. Ajmal Khan, P. Sichanugrist, S. Kato, and Y. Ishikawa: Energy. Sci. Eng. **4** (2016) 383.

[53] M. Ajmal Khan R. Sato, K. Sawano, P. Sichanugrist, A. Lukianov, and Y. Ishikawa: J. Phys. D: Appl. Phys. **51** (2018) 185107.

[54] Y. C. Kim, S. J. Lee, and J.-M. Myoung: J. Phys. Chem. C **121** (2017) 6788.

Chapter 8

Nanowires Integrated to Optical Waveguides

Ricardo Téllez-Limón and Rafael Salas-Montiel

Abstract

Chip-scale integrated optical devices are one of the most developed research subjects in last years. These devices serve as a bridge to overcome size mismatch between diffraction-limited bulk optics and nanoscale photonic devices. They have been employed to develop many on-chip applications, such as integrated light sources, polarizers, optical filters, and even biosensing devices. Among these integrated systems can be found the so-called hybrid photonic-plasmonic devices, structures that integrate plasmonic metamaterials on top of optical waveguides, leading to outstanding physical phenomena. In this contribution, we present a comprehensive study of the design of hybrid photonic-plasmonic systems consisting of periodic arrays of metallic nanowires integrated on top of dielectric waveguides. Based on numerical simulations, we explain the physics of these structures and analyze light coupling between plasmonic resonances in the nanowires and the photonic modes of the waveguides below them. With this chapter we pretend to attract the interest of research community in the development of integrated hybrid photonic-plasmonic devices, especially light interaction between guided photonic modes and plasmonic resonances in metallic nanowires.

Keywords: plasmonics, integrated optics, nanowires, optical waveguides, hybrid modes

1. Introduction

Plasmonics, the science of plasmons, is a research field that has been extensively studied in recent years due to its multiple applications like biosensing, optical communications, or quantum computing, to mention but a few.

Generally, the field of plasmonics is associated with two types of collective oscillations of conductive electrons at the boundaries of metallic nanostructures, known as surface plasmon polaritons (SPP) and localized surface plasmons (LSP). While SPP are referred as surface waves propagating at a dielectric-metal interface, LSP can be regarded as standing surface waves confined in metallic nanoparticles embedded in a dielectric environment [1].

As it is well known, SPP modes can only be excited when appropriate phase match conditions are fulfilled. An option to achieve this condition, is by making use of the electromagnetic near field scattered by a local defect or emitter. To this purpose, the LSP mode of a metallic nanoparticle can be excited and coupled to the SPP of a metallic substrate, giving rise to hybrid plasmon polaritons [2, 3].

In addition to these types of plasmonic oscillations, there are other resonances named plasmonic chain modes. These modes can be generated in linear arrays of closely spaced metallic nanoparticles, including nanowires, and they result from the near field coupling between adjacent nanoparticles excited at their plasmonic resonances. Due to this coupling effect, light can propagate through the periodic arrays. Thus, these periodic structures can be regarded as discrete plasmonic waveguides [4–6]. When placing a periodic array of metallic nanoparticles in a layered media, under proper excitation conditions, the plasmonic chain modes can also couple to the SPP of a metallic substrate, forming hybrid SPP-chain modes [7].

In this same sense, when placing periodic arrays of metallic nanoparticles on top of dielectric waveguides, the plasmonic chain modes can couple to the photonic modes of the waveguide [8]. These integrated structures give rise to the so-called hybrid photonic-plasmonic waveguide modes [9], and they are the main subject of interest in this chapter. We will focus our attention to integrated structures consisting of periodic arrays of metallic nanowires integrated on top of two-dimensional dielectric photonic waveguides.

To this purpose, we will bring a comprehensive explanation about the physics behind the dispersion curves of integrated hybrid photonic-plasmonic waveguiding structures. Then will be studied the propagation of electromagnetic fields through the integrated systems varying the geometric cross-section of the metallic nanowires. For a better understanding, this comprehensive study will be accompanied by numerical simulations, making easier to elucidate the potential applications of these outstanding structures.

2. Hybrid photonic-plasmonic waveguides

2.1 Optical waveguides

From the analysis of the chemical composition of farer stars to imaging of microscopic living cells, information transport through light is one of the main subjects of interest in optical sciences. Among the different ways to transport light can be found optical waveguides, whose principle of operation is based on the total internal reflection effect. This phenomenon consists of the complete reflection of light within a medium surrounded by media with smaller refractive index, as depicted in **Figure 1**.

The schematic in **Figure 1a** represents an asymmetric planar waveguide invariant along the out-of-plane direction, consisting of a dielectric medium of refractive index n_2 between two media of refractive index n_1 and n_3, where $n_2 > n_1 > n_3$. As light propagates in the inner medium n_2, certain rays will present a phase difference of zero or a multiple of 2π, when they are twice reflected. This situation means that

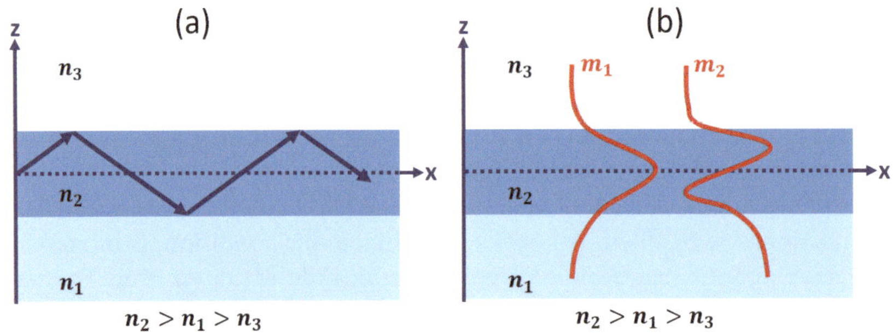

Figure 1.
Schematic representation of a planar asymmetric waveguide consisting of three dielectric media of refractive index n_1, n_2 and n_3. (a) Self-consistency condition defining the modes of the waveguide. (b) Profile of the field distributions of the first two guided modes of the waveguide.

after a round-trip the wave reproduces itself, preserving the same spatial distribution and polarization along the waveguide. Fields satisfying this self-consistency condition are known as eigenmodes or modes of the waveguide [10]. The schematic in **Figure 1b** shows the profile of the field distribution for the first two modes of the proposed waveguide.

2.2 Dispersion relation

To determine the propagation constant of the modes supported by the waveguide, let us consider a waveguide with a core of refractive index n_2 and thickness d, surrounded by two semi-infinite dielectric media of refractive index n_1 and n_3, as depicted in **Figure 2**.

For each medium, the field can be represented as a sum of propagative and counter-propagative waves along the z axis, and propagative in the x direction that can be represented as

$$\psi_m(x, z, \omega) = A_m e^{-i\alpha_m z} e^{-i\beta_m x} e^{-i\omega t} + B_m e^{i\alpha_m z} e^{-i\beta_m x} e^{-i\omega t}, \tag{1}$$

where $m = I, II, III$, A_m and B_m are the amplitudes of the propagative and counter-propagative waves, respectively, and the propagation constants α_m and β_m along the z and x axis are related through

$$\beta_m = \sqrt{\left(\frac{\omega}{c}\right)^2 \varepsilon_m(\omega) - \alpha_m^2}, \tag{2}$$

where $\varepsilon_m(\omega)$ is the dielectric constant of the m-th medium related to the refractive index by $n_m = \sqrt{\varepsilon(\omega)\mu(\omega)}$. At optical wavelengths, he magnetic permeability $\mu(\omega)$ can be considered as unit. Req. (2) is obtained from the Helmholtz and Maxwell equations [11].

At the interfaces $z = 0$ and $z = d$, the electromagnetic field should be continuum, that is to say:

$$\psi_I(x, z, \omega)\big|_{z=0} = \psi_{II}(x, z, \omega)\big|_{z=0}, \tag{3}$$

and

$$\psi_{II}(x, z, \omega)\big|_{z=d} = \psi_{III}(x, z, \omega)\big|_{z=d}. \tag{4}$$

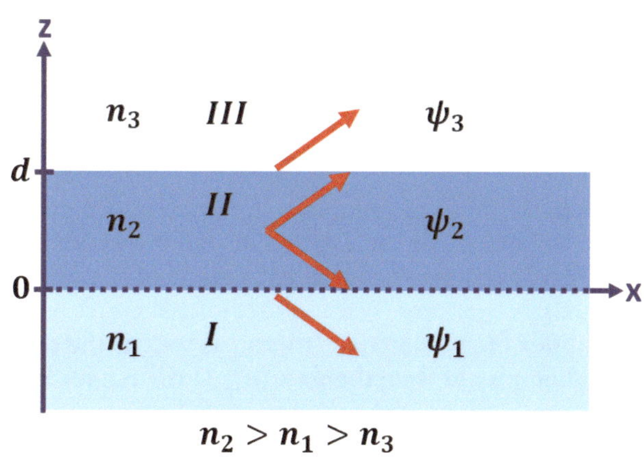

Figure 2.
Schematic representation of the field components in an asymmetric planar waveguide.

From the conservation of the tangential components of the electromagnetic field at the boundaries between two media [12] are obtained the relationships

$$\frac{1}{\nu_I}\frac{\partial\psi_I(x,z,\omega)}{\partial z}\bigg|_{z=0}=\frac{1}{\nu_{II}}\frac{\partial\psi_{II}(x,z,\omega)}{\partial z}\bigg|_{z=0}, \tag{5}$$

$$\frac{1}{\nu_{II}}\frac{\partial\psi_{II}(x,z,\omega)}{\partial z}\bigg|_{z=-d}=\frac{1}{\nu_{III}}\frac{\partial\psi_{III}(x,z,\omega)}{\partial z}\bigg|_{z=d}, \tag{6}$$

with $\nu_m = 1$ for TE polarized electromagnetic fields and $\nu_m = \varepsilon_m(\omega)$ for TM polarized fields. Substituting Eq. (1) in Eqs. (3–6), and considering that $A_I = B_{III} = 0$ because both, I and III are semi-infinite media and no back-reflections from the boundaries are present, it is obtained a two coupled equation system that can be represented in a matrix way of the form

$$\begin{bmatrix} \frac{\alpha_I}{\nu_I}-\frac{\alpha_{II}}{\nu_{II}} & \frac{\alpha_I}{\nu_I}+\frac{\alpha_{II}}{\nu_{II}} \\ e^{-i\alpha_{II}d}\left(\frac{\alpha_I}{\nu_I}+\frac{\alpha_{III}}{\nu_{III}}\right) & -e^{-i\alpha_{II}d}\left(\frac{\alpha_I}{\nu_I}-\frac{\alpha_{III}}{\nu_{III}}\right) \end{bmatrix}\begin{bmatrix} A_{II} \\ B_{II} \end{bmatrix}=\begin{bmatrix} 0 \\ 0 \end{bmatrix}. \tag{7}$$

By equating to zero the determinant of the matrix it is possible to obtain the non-trivial solutions of this eigenmode equation system, resulting in the dispersion relation of a three-layered media

$$\frac{\left(\frac{\alpha_{II}}{\nu_{II}}-\frac{\alpha_I}{\nu_I}\right)\left(\frac{\alpha_{II}}{\nu_{II}}-\frac{\alpha_{III}}{\nu_{III}}\right)}{\left(\frac{\alpha_{II}}{\nu_{II}}+\frac{\alpha_I}{\nu_I}\right)\left(\frac{\alpha_{II}}{\nu_{II}}+\frac{\alpha_{III}}{\nu_{III}}\right)}=e^{i2d\alpha_{II}}. \tag{8}$$

We must notice that Eq. (8) is a transcendental function with no analytical solution, thus, numerical methods should be employed to solve it.

When solving this dispersion relation, it is obtained the mode propagation constant, β, that depends on the optical frequency or wavelength of light and determines how the amplitude and phase of light varies along the x direction. In the same way as wavenumber can be related to the refractive index of a homogeneous medium, the propagation constant can be regarded as the wavenumber (spatial frequency) of light propagating through an effective medium composed by the inhomogeneous three-layered structure. The propagation constant is then related to the so-called effective index through the relationship

$$\beta=\frac{2\pi}{\lambda}n_{eff}, \tag{9}$$

being λ the wavelength of light in vacuum. We must notice that the effective index is only defined for a mode of the waveguide and it should not be understood as a material property. We can say then that each mode of the waveguide will "see" different effective media.

As the refractive index of a dielectric medium, as well as the dielectric constant, is a real number equal or greater than the unit ($n \geq 1$) the modes in a dielectric waveguide are diffraction limited: if the thickness of the waveguide, d, is smaller than $\lambda/(2n_{eff})$, the solutions for the dispersion Eq. (8) will lead to evanescent waves, meaning that no modes can be propagated below this limit.

2.3 Plasmonic waveguides

As previously explained, dielectric waveguides guide light modes by using the total internal reflection principle and self-consistency condition. These waveguides are diffraction limited due to the dielectric constant values. However, if the dielectric constant is a complex number, it would be possible to obtain solutions to the dispersion relation (Eq. 8) below the diffraction limit. This is the case of metallic materials. Hence, if at least one of the three media in the waveguide structure is metallic, it is always possible to obtain a propagative mode in the structure. The price to pay for this solution is that due to ohmic losses in metals, these modes propagate just few microns, in opposition to dielectric waveguides where light can propagate through kilometers.

These structures are known as plasmonic waveguides, and their operation principle is based on SPP mode propagation. These surface waves are the result of collective oscillations of the conductive electrons at a metal-dielectric interface induced by the electric field of an electromagnetic wave. For a system invariant in the $\hat{y}$ direction, SPP modes can only be excited if the electric field oscillates in the xz plane. Hence, only TM polarized electromagnetic fields couple to SPP modes (for TE polarized waves the electric field only oscillates along the $\hat{y}$ direction).

Different combinations of insulator (I) and metallic (M) materials can be used to define a plasmonic waveguide. In **Figure 3** are represented IIM, IMI and MIM plasmonic waveguide structures as well as the amplitude distribution of the out-of-plane electromagnetic field (H_y component) of the SPP modes. For the IIM structure, there is only one SPP mode at the interface between the metal (ε_m) and first dielectric (ε_{d_1}). For both IMI and MIM configurations, two SPP modes can be excited. They result from in-phase and out-of-phase coupling of SPP at the first and second dielectric-metal interfaces, and they are known as symmetric and antisymmetric modes, respectively.

As plasmonic waveguides allow light propagation beyond the diffraction limit, these structures have been used for the development of integrated nanophotonic devices for optical signal transportation, optical communications, biosensing and even imaging applications [13–15].

2.4 Hybrid photonic-plasmonic waveguides

From the previous waveguiding configurations, it is natural to think that modes propagating through a dielectric waveguide can be coupled to a plasmonic waveguide. This kind of structures is named hybrid photonic-plasmonic waveguide, or simply hybrid plasmonic waveguide.

The structure depicted in **Figure 3a** can be considered as a hybrid plasmonic waveguide, but more complex multilayered systems can be designed to propagate more than one mode in these structures. For instance, in **Figure 4** are presented two

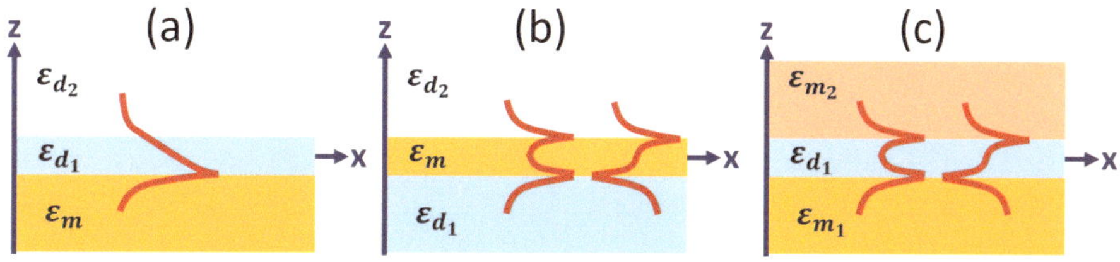

Figure 3.
Schematic representation of plasmonic waveguides for (a) IIM, (b) IMI and (c) MIM configurations and SPP modes profiles. For IMI and MIM waveguides, symmetric and antisymmetric modes are excited.

examples of hybrid plasmonic waveguides able to support symmetric and antisymmetric SPP modes coupled to photonic modes of a dielectric waveguide.

To compute the supported modes of these structures, we can make use of the dispersion relation for a N-layered medium in terms of the T-matrix that relates the amplitudes of propagative and counter-propagative waves, A_m and B_m, in the m-th medium, to those from the $m+1$ medium through the relationship [16].

$$\begin{bmatrix} A_{N+1} \\ B_{N+1} \end{bmatrix} = [\mathbb{T}] \begin{bmatrix} A_1 \\ B_1 \end{bmatrix} = \prod\nolimits_{m=1}^{N} [T_m] \begin{bmatrix} A_1 \\ B_1 \end{bmatrix} = \begin{bmatrix} t_{11} & t_{12} \\ t_{21} & t_{22} \end{bmatrix} \begin{bmatrix} A_1 \\ B_1 \end{bmatrix}, \tag{10}$$

where $A_1 = B_{N+1} = 0$ (no back reflections from substrate and superstrate) and

$$[T_m] = \frac{1}{2} \begin{bmatrix} (1+\gamma)e^{-i(k_m - k_{m+1})} & (1-\gamma)e^{i(k_m + k_{m+1})} \\ (1-\gamma)e^{-i(k_m + k_{m+1})} & (1+\gamma)e^{i(k_m - k_{m+1})} \end{bmatrix}, \tag{11}$$

with $k_m = \alpha_m d_m$, $k_{m+1} = \alpha_{m+1} d_m$, $\gamma = (\alpha_m \nu_{m+1})/(\alpha_{m+1}\nu_m)$, being d_j the position of the interface between j and $j+1$ media, considering that $d_{N+1} = d_N$ and $\nu_m = 1$ for TE polarized electromagnetic fields and $\nu_m = \varepsilon_m(\omega)$ for TM polarized fields. By equating to zero the term t_{22} of the $\mathbb{T}$ matrix, we can directly obtain the propagation constant of the modes supported by the structure.

2.5 Dispersion curves and mode analysis

Before studying light propagation in complex hybrid plasmonic waveguides, it is worthily to briefly comment on the dispersion diagrams that would help to perform an analysis of the modes propagating through these waveguides. So far, we have presented the dispersion relation for a multilayered media. It is not our intention to explore the numerical methods that can be employed to solve this transcendental equation, but to analyze the information obtained from these results. The reader can look at references [17–22] to have an insight of how to solve the dispersion relation.

As an example, let us analyze the modes of a four-layered media as schematized in **Figure 4a**, consisting of a glass substrate with refractive index $n_1 = 1.5$, a silicon nitride layer (core of the photonic waveguide) of thickness $d = 300$ nm and refractive index $n_c = 2.0$, a thin gold layer of thickness $t = 40$ nm, and air superstrate ($n_2 = 1.0$). The numerical results obtained from the calculation of the dispersion relation for TM polarized fields by using the Raphson–Newton method [23] are plotted in **Figure 5**. For these calculations was considered a spectral wavelength range from 400 nm to 1 μm, and effective index range between 0.9 and 2.5. Since

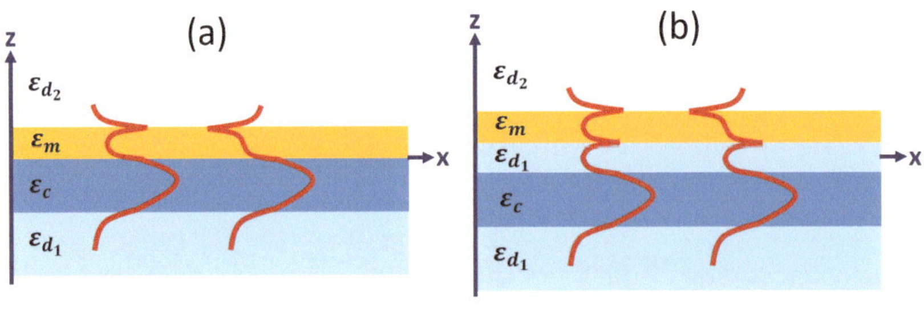

Figure 4.
Schematic representation of hybrid photonic-plasmonic waveguides and mode profiles for (a) a metallic layer placed directly on top of the dielectric waveguide, and (b) with an intermediate dielectric layer between photonic and plasmonic waveguides. In both systems, the fundamental mode of the waveguide couple to the symmetric and antisymmetric SPP modes.

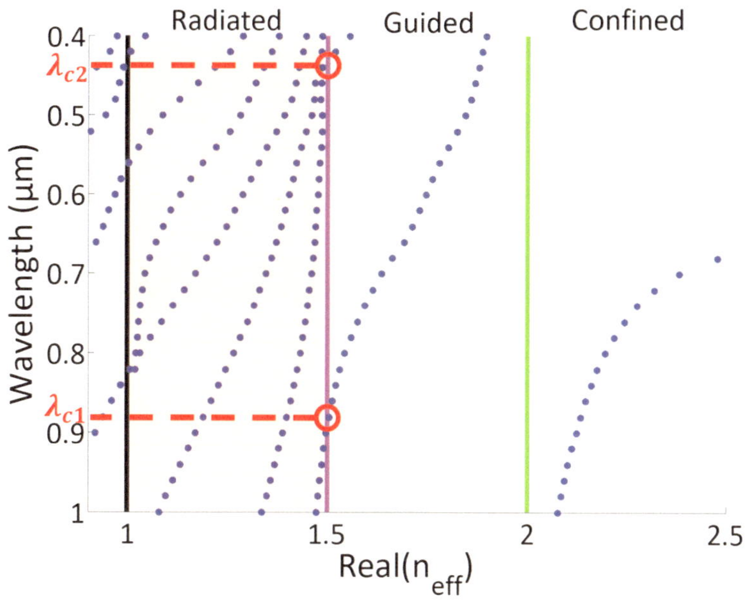

Figure 5.
Dispersion curves for a hybrid photonic-plasmonic waveguide consisting of a glass substrate ($n_1 = 1.5$), silicon nitride core of 300 nm thickness ($n_c = 2.0$), a thin gold layer of thickness 40 nm, and an air superstrate ($n_2 = 1.0$). The numerical results show a confined mode (blue region) and two guided modes (green region). At the orange region, many radiated modes were obtained, as well as many non-physical solutions (gray region).

the results are just numerical solutions, we need to understand the physical meaning for each solution.

The vertical constant lines at 1, 1.5 and 2.0 correspond to the refractive index of each dielectric medium: air superstrate, glass substrate and silicon nitride core, respectively. These vertical lines are also referred as light lines, as they are linked to the propagation constant of light traveling in that specific homogeneous medium through Eq. (9).

These light lines define four different regions. The first region, for effective index values below 1 (gray region), are numerical solutions without physical meaning: if the effective index is smaller than unit, the modes would travel faster than speed of light in vacuum (which obviously is not our case). The second region between the refractive index of glass and air refractive index (orange region) defines modes with effective index smaller than glass but greater than air. Hence, they are modes whose energy is propagating in the glass substrate, and they are referred as radiated modes. The third region, between the silicon nitride (core) light line and glass light line, define modes whose energy is propagating in the core of the waveguide: as the effective index is higher than glass substrate index, the energy of these modes does not propagates in glass, so the energy is confined in the core. These are guided modes. The value at which the effective index of these modes matches the refractive index of the glass substrate determines the cut-off wavelength of guided modes. For the analyzed example, these values are $\lambda_{c1} = 886$ nm and $\lambda_{c2} = 430$ nm (red circles).

The fourth region (blue colored) correspond to modes whose energy does not propagates in any of the dielectric layers: their effective index is greater than the core, substrate, and superstrate. Hence, these modes are confined to the metallic layer. These solutions correspond to propagative SPP modes and they are referred as confined modes.

In literature, different representations of the dispersion curves can be found, like propagation constant vs. frequency (usually normalized to a reference value), wavelength vs. incidence angle (used in attenuated total internal reflection measurements), among others. The representation that we use in **Figure 5** allows us to understand the dispersion curves in terms of two quantities that can be easily identified: wavelength and effective index.

2.6 Mode hybridization

To further understand the origin of the modes appearing in the hybrid photonic-plasmonic structure, let us analyze the multilayered system by parts: first we will compute the dispersion curves of the photonic waveguide (**Figure 6a**), then the modes of the plasmonic waveguide (**Figure 6b**), and finally compare them with the full hybrid photonic-plasmonic structure (**Figure 6c**). The numerical results for the dispersion relation of each one of these cases are presented in **Figure 6d**. The dimensions of the structures are the same than those used for **Figure 5**.

The blue dots and circles in guided region, correspond to the fundamental TM_0 and higher order TM_1 modes of the silicon nitride waveguide (thickness $d = 300$ nm and refractive index $n_c = 2.0$) surrounded by air superstrate ($n_{sup} = 1.0$) and glass substrate ($n_{sub} = 1.5$), as depicted in **Figure 6a**. These modes present cut-off wavelength values around $\lambda_{TM0} = 1.58$ μm and $\lambda_{TM1} = 545$ nm, respectively.

When computing the modes of a thin gold layer of thickness $t = 40$ nm on top of a glass substrate ($n_{sub} = 1.5$) and air superstrate, as depicted in **Figure 6b**, it is observed one mode in the guided region that tends to a constant value (red dots in **Figure 6d**). For this structure, we must notice that no core was present, then, the effective index of this mode is higher than the refractive index of the glass substrate, hence, it is a SPP mode confined to the metallic layer.

We can observe that both, TM_0 and SPP modes, cross each other around an effective index value of 1.798. In other words, the propagation constant for both modes are the same, so they are phase matched. This situation means that the photonic mode of the waveguide will excite the plasmonic mode of the metallic layer.

When the modes of the complete integrated structure (**Figure 6c**) are computed, two branches are observed. The first one, represented by red triangles, is a

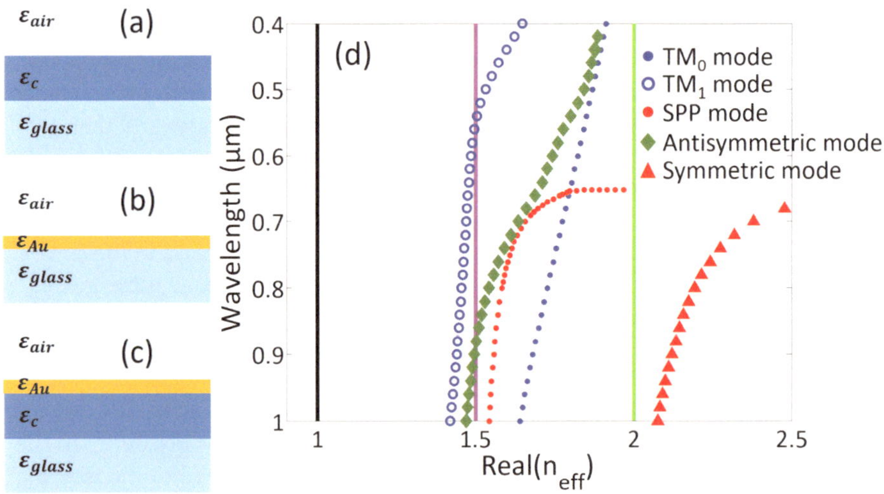

Figure 6.
Schematic representation of (a) photonic waveguide, (b) plasmonic waveguide, and (c) hybrid photonic-plasmonic waveguide. (d) Dispersion curves for the three studied structures. Due to phase matching between TM_0 and SPP modes, hybrid symmetric and antisymmetric modes arise in the full integrated structure.

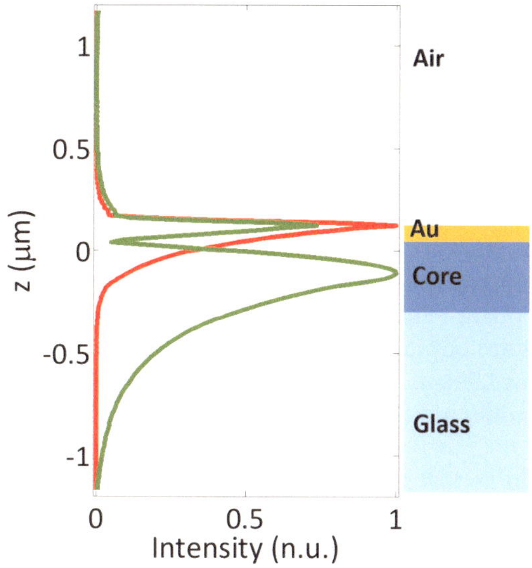

Figure 7.
Intensity profiles of the symmetric (red) and antisymmetric (green) modes of the hybrid photonic plasmonic waveguide at $\lambda = 750$ nm. The symmetric mode is mainly confined in the metallic layer, while the antisymmetric presents amplitude in both, photonic and plasmonic waveguides.

mode confined to the metallic layer, and the second one, relies in the guided region (green diamonds). These modes arise from the coupling of the TM_0 and SPP modes, and they are referred as hybrid modes, being the first one a symmetric mode and the second an antisymmetric mode.

It is important to say that both symmetric and antisymmetric modes are not independent, they are hybrid modes. Like in two coupled harmonic oscillators, this hybridization means that there is an energy exchange between photonic and plasmonic waveguides. For the symmetric mode, the amplitude of light in both, photonic and plasmonic waveguides, are in phase, while for the antisymmetric are out-of-phase [24, 25].

Finally, in **Figure 7** are plotted the normalized intensity profiles of the symmetric (red curve) and antisymmetric (green curve) modes at a wavelength of $\lambda = 750$ nm. The intensity is derived from the amplitudes of Eq. 1. As expected from the dispersion curves, the intensity of the symmetric mode is mainly confined in the metallic layer (the mode solution relies in the confined region), while for the antisymmetric mode the intensity is distributed in both photonic and plasmonic waveguides, being greater the intensity in the dielectric region (the solution relies in the guided region).

3. Mode propagation in a periodic array of metallic nanowires

In general, plasmonic resonances in metallics nanostructures are divided in two kinds, namely SPP and LSP. SPP modes are propagative waves confined at the dielectric/metal interface, while LSP are standing waves or cavity modes oscillating in a nanoparticle.

As it is well known, LSP resonances depend not only on the material of the nanoparticles, but also on their shape and polarization of the incident wave: the orientation of the electric field defines the direction of the oscillation of the charges in the metallic nanoparticle; these charges will distribute depending on the geometry of the particle, giving rise to different modes. For small nanoparticles, usually are only excited dipolar LSP resonances, but quadrupoles, octupoles and higher order modes can also be excited.

When metallic nanoparticles are closely placed and excited at their LSP resonance, it is possible to couple them via near field interaction, leading to higher order LSP modes. To understand this coupling mechanism, let us take a look to **Figure 8**, where a dimmer of spherical metallic nanoparticles oriented along the x axis (dimmer axis), is excited with electric field oscillating in z and x directions.

When the electric field is oriented along the z axis, perpendicular to the dimer axis, the dipolar resonances of the nanoparticles are oriented also in the z direction. If the dipoles are in phase (**Figure 8a**), the dimmer also presents a dipolar resonance. If the dipoles are out-of-phase (**Figure 8b**), the dimmer presents a quadrupolar resonance. Since the distribution of the charges is perpendicular to the dimmer axis, both modes are referred as dipolar and quadrupolar transverse modes, respectively.

If the electric field oscillates along the x axis, the dipoles of the nanoparticles will be oriented along the dimmer axis, thus, the coupled modes are called dipolar longitudinal modes. If the dipoles are in phase (**Figure 8c**), the resonance wavelength of the longitudinal mode will be shorter than the resonance wavelength of the out-of-phase dipoles (**Figure 8d**).

In the same way, a periodic array of metallic nanoparticles can be coupled, allowing light propagation. Thus, when properly excited, a periodic array of metallic nanoparticles can be regarded as a plasmonic waveguide. These resonances are named plasmonic chain modes, and their waveguiding properties will depend on the shape and period of the nanoparticles, as well as the orientation of the incident electromagnetic field. Besides energy transportation capabilities, these modes have been widely studied because they allow a strong enhancement of the electromagnetic field in a localized nanometric region.

If we consider that the nanoparticles in **Figure 8** are invariant in the out-of-plane y direction, the same coupling mechanism is preserved. These structures are named metallic nanowires (MNW): metallic nanostructures with nanometric cross section and micrometric lengths. As the length of the nanowires is times longer than the incident wavelength, the absorption of light prevents from the formation of cavity modes in this direction. This means that plasmonic resonances in metallic nanowires can be excited only if the electric field is perpendicular to the invariant

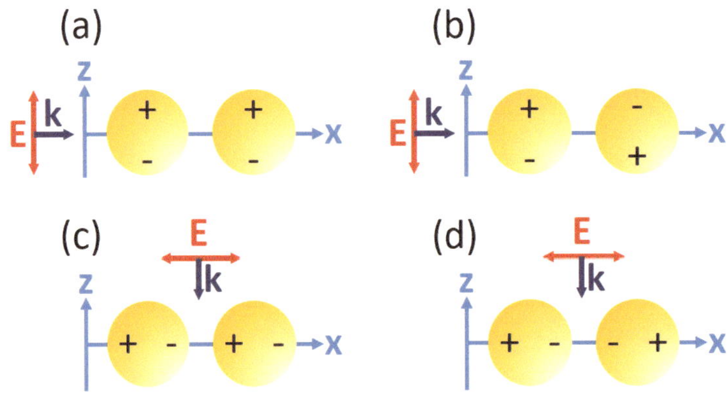

Figure 8.
LSP coupled modes for spherical nanoparticle dimmers. When the electric field is perpendicular to the main axis of the dimmer are excited (a) dipolar and (b) quadrupolar transverse modes. When the electric field is parallel to the main axis of the dimmer, dipolar longitudinal modes of (c) shorter and (d) longer resonant wavelengths are obtained.

y axis, i. e. with TM polarized light. Thus, the plasmonic chain modes in metallic nanowires will mainly depend on the geometry of their cross section [26–28].

3.1 Plasmonic chain modes in MNW with rectangular cross section

Let us consider an infinite periodic array of gold nanowires of width $w = 80$ nm, gap between them $g = 50$ nm (period $\Lambda = 130$ nm) and thickness $t = 20$ nm, embedded in a homogeneous dielectric medium of refractive index $n_d = 1.5$ (glass), as depicted in **Figure 9a**. The structure in invariant along the out-of-plane direction.

To perform a modal analysis, it is required to compute the dispersion curves of these system. Different numerical methods can be employed, like effective index method [29–31], source model technique [32], rigorous coupled wave analysis (RCWA) [33–35], or Fourier modal method (FMM) [36–39], among others.

In our case, we will make use of the FMM to compute the dispersion curves of the periodic structure. This rigorous method computes the Maxwell equations in the frequency domain. To solve them, a unit cell of the periodic structure, as well as the dielectric function and electromagnetic field are expanded in Fourier series. This formulation leads to an eigenvalue matrix formulation that can be used to obtain the modes of the nanowires in a multilayered media. Also, by adding perfectly matched layers (PML), it is possible to compute the beam propagation in a finite periodic structure. It is not our intention to show how to implement this numerical method, but to analyze plasmonic chain modes in periodic MNW. For a better comprehension about this method, we invite the reader to look at references [37, 38].

The plot in **Figure 9b** corresponds to the dispersion curves of the system under study. As we are dealing with a periodic structure, it is useful to represent this plot in terms of the propagation constant (along the periodicity direction), normalized to the Bragg condition ($\beta = \pi/\Lambda$), which defines the first Brillouin zone (vertical black line). The red dotted curve represents the modes supported by the MNW, as they are confined below the glass light-line. To understand the behavior of this

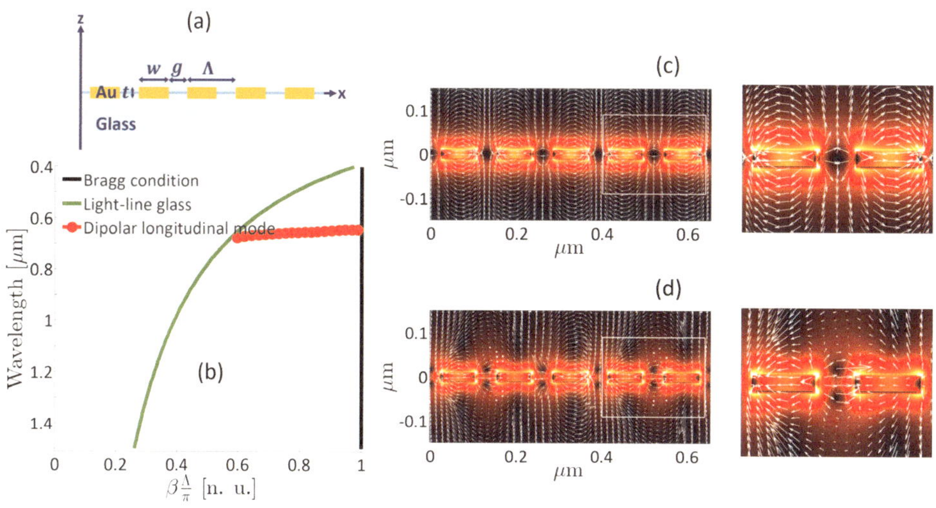

Figure 9.
(a) Schematic representation of a periodic infinite array of gold nanowires of width $w = 80$ nm, gap $g = 50$ nm, period $\Lambda = 130$ nm and thickness $t = 20$ nm in a homogeneous dielectric medium. (b) the confined modes below the glass light-line (green curve) correspond to a dipolar longitudinal mode (red dotted curve). (c) the energy density map and electric field lines computed at the Bragg condition ($\lambda = 645$ nm), reveals the dipolar longitudinal coupling between MNW. (d) out of the Bragg condition, at $\lambda = 667$ nm, we still observe the dipolar coupling.

plasmonic mode, it is necessary to observe the distribution of the charges in the MNW.

In **Figure 9c** are shown the energy density map and electric field lines distribution of the plasmonic mode at $\lambda = 645$ nm. At this wavelength, the red curve crosses the Bragg condition, defining a stationary mode. As we can observe, MNW are coupled, with a phase shift of π rad between them. Thus, it is a plasmonic chain mode. Out of the Bragg condition, for instance, at $\lambda = 667$ nm ($\beta = 0.641\pi/\Lambda$), the chain mode becomes propagative and the electric field lines remain almost longitudinally oriented inside the MNW (**Figure 9d**). In view of the phase shift and the orientation of the electric field lines, the plasmonic chain mode results from coupled dipolar resonances oriented along the x axis. Thus, the red dotted curve corresponds to the dispersion relation of a dipolar longitudinal plasmonic chain mode.

As defined for a SPP, the propagation length of this plasmonic chain mode can be computed through the relationship

$$L_p = \frac{1}{2\beta''}, \tag{12}$$

where β'' is the imaginary part of the propagation constant. In our structure, the propagation length varies from $L_p = 200$ nm (for wavevectors close to the glass light-line) up to $L_p = 1.14$ μm (for wavevectors near the Bragg condition).

3.2 Hybrid plasmonic chain modes in MNW integrated to a dielectric waveguide

Now, let us study a hybrid photonic-plasmonic system consisting of a dielectric waveguide of thickness $d = 200$ and refractive index $n_{wg} = 2.0$, on top of which is placed, at a distance $h = 30$ nm, a periodic array of gold nanowires with the same parameters than those of the previous subsection, as depicted in **Figure 10a**.

The dispersion curves in the first Brillouin zone, obtained with the FMM, are shown in **Figure 10b**. The green and magenta curves represent the glass and waveguide light lines, respectively, while the black vertical line represents the Bragg condition. The blue triangles curve corresponds to the fundamental TM_0 photonic guided mode without the presence of the MNW, while the green asterisk curve is the dispersion curve of the dipolar longitudinal plasmonic chain mode (without the presence of the waveguide). As can be observed, there is an intersection point between the TM_0 and dipolar longitudinal modes in the guided region around a wavelength value $\lambda = 664$ nm (**Figure 10c**). This crossing point means that both modes have the same propagation constant ($\beta = 0.672(\pi/\Lambda)$), hence, they will couple when placing them near to each other, leading to a hybrid photonic-plasmonic mode. As a result, symmetric and anti-symmetric modes will arise. This hybridization is corroborated when computing the dispersion curves of the integrated system, being observed two curves: a lower branch passing from guided to confined region (red circles) and an upper branch in the guided region. In analogy with a two coupled waveguide system, this situation means that, for the lower branch, energy from the TM_0 mode is converted into plasmonic chain mode; for the upper branch, the energy from the plasmonic chain mode is converted into TM_0 mode. This anti-crossing phenomenon is the characteristic signature of strong coupling between guided modes.

To corroborate the symmetry of these modes, in **Figure 11** we plot the energy density maps and electric field lines distribution for both modes at the Bragg condition. **Figure 11a** corresponds to the energy density map computed for the upper branch at $\lambda = 473$ nm, where we can observe that the energy is distributed in

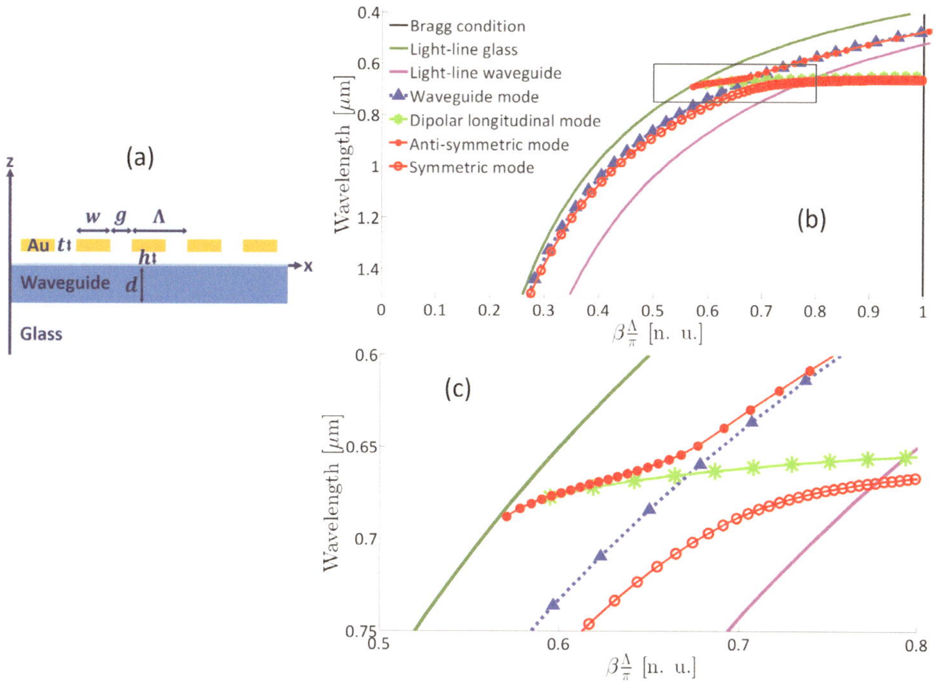

Figure 10.
(a) Schematic representation of the integrated structure consisting of an infinite periodic arrays of gold nanowires ($w = 80$ nm, $g = 50$ nm, $\Lambda = 130$ nm, $t = 20$ nm) surrounded by glass ($n_s = 1.5$), placed on top of a dielectric waveguide ($d = 200$ nm, $n_{wg} = 2.0$) at a distance $h = 30$ nm. (b) Dispersion curves of the integrated system. (c) Inset at the anti-crossing region.

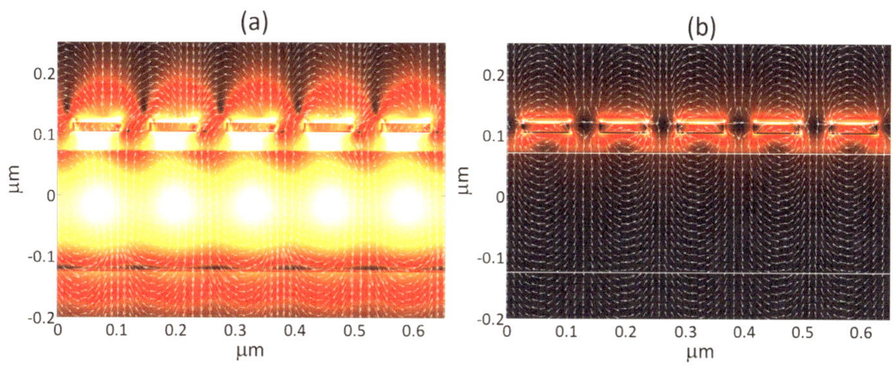

Figure 11.
Energy density maps and electric field lines distribution for (a) the antisymmetric mode at $\lambda = 473$ nm, and (b) for the symmetric mode at $\lambda = 660$ nm.

the dielectric waveguide and the periodic array of MNW. When looking at the electric field lines, it is observed a phase difference between these vectors above and below the MNW, defining an antisymmetric quadrupolar chain mode. For the case of the lower branch, computed at $\lambda = 660$ nm (**Figure 11b**), the energy is confined to the MNW and not in the waveguide, as expected from the dispersion curves. For this case, the electric field lines above and below the MNW are in phase, defining a symmetric dipolar chain mode.

Finally, by making use of the aperiodic Fourier modal method [27, 39], and considering a finite number of 27 MNW, we computed the transmission (red curve), reflection (blue dashed curve), and absorption (black dotted curve) spectra of light propagating through the integrated system, normalized to the incident beam (**Figure 12a**). The reflection spectrum (blue dashed curve) exhibits a maximum peak around $\lambda = 474$ nm, while the transmission spectrum shows a broad depth around $\lambda = 520$ nm and a narrower depth around $\lambda = 678$ nm. The first depth (close to the maximum in reflection) is due to a Bragg reflection induced by the periodic array of MNW. This condition can be verified with the expression

$$\lambda_{Bragg} = \frac{2n_{eff}\Lambda}{m}, \tag{13}$$

where n_{eff} is the effective index of the mode, Λ the period of the MNW array, and m the Bragg order. According to the dispersion curves, at $\lambda = 520$ nm, the effective index of the antisymmetric mode is $n_{eff} = 1.801$, and considering the period $\Lambda = 130$ nm, the first order Bragg reflection will occur at $\lambda_{Bragg} = 468$ nm, which is a value close to the maximum in the reflection spectrum. The near field map in **Figure 12b**, shows the amplitude of the H_y field component, where are observed periodic lobes inside the waveguide core due to the Bragg reflection.

The second depth on the transmission spectrum corresponds to the excitation of the dipolar longitudinal plasmonic chain mode. In the dispersion curves, this wavelength value corresponds to the anti-crossing point between symmetric and antisymmetric modes, around $\beta = 0.672(\pi/\Lambda)$. As expected, at this wavelength the plasmonic chain mode is efficiently coupled to the photonic TM_0 mode of the waveguide, leading to an anergy exchange between the photonic waveguide and the periodic array of MNW, as illustrated in the near field map of **Figure 12c**.

3.3 Hybrid plasmonic chain modes in MNW of triangular cross section integrated to a dielectric waveguide

Among the large variety of shapes in nanowires, sharp geometries such as nanotips stimulate a great interest in applications where a strong localization of the electromagnetic field is required. These triangular geometries present an

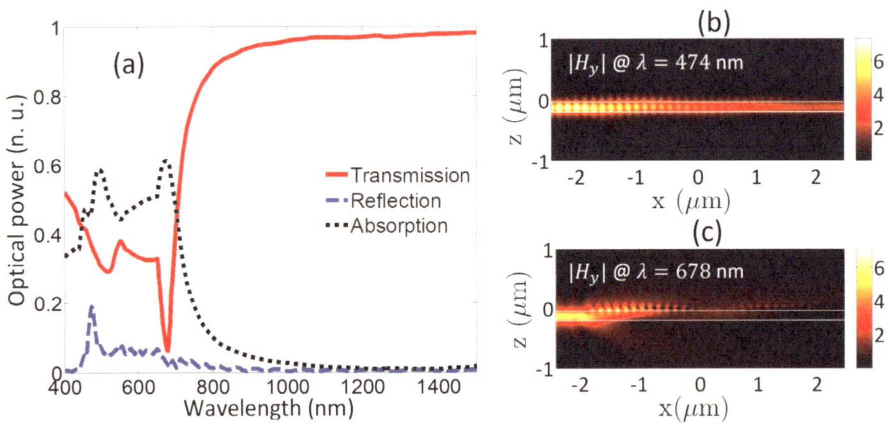

Figure 12.
(a) Normalized transmission (red), reflection (blue dashed) and absorption (black dotted) spectra of light propagating through the waveguide in the integrated device. (b) At $\lambda = 474$ nm, the H_y field component exhibits a Bragg reflection, while at (c) $\lambda = 678$ nm, the it is observed an energy exchange between the waveguide and periodic array of MNW.

extraordinary enhancement of light in the vicinity of their apex resulting from the excitation of LSP resonances polarized along their tip axis. In this section, we will study the excitation of plasmonic chain modes in periodic arrays of gold nanowires with triangular cross section through the photonic mode of a dielectric waveguide.

Firstly, we will study the plasmonic chain modes of the MNW placed on top of a glass substrate and the photonic modes of the dielectric waveguide that will be used to excite them. The MNW consist of an infinite periodic array of gold nanowires with triangular cross section of height $t = 144$ nm, width $w = 72$ nm, and period $\Lambda = 200$ nm, placed on top of a glass substrate ($n_{sub} = 1.5$). The superstrate is air ($n_{sup} = 1.0$), the tip radius of the nanowires is $r = 5$ nm, and the system is invariant in the out-of-plane y direction, as depicted in **Figure 13a**. Then dielectric waveguide consist of a core of thickness $d = 200$ nm end refractive index $n_{wg} = 2.0$, buried a distance $h = 30$ nm from the glass/air interface (**Figure 13b**).

The dispersion curves in the first Brillouin zone are plotted in **Figure 13c**, where can be observed the light lines of air superstrate (white curve), glass substrate (black curve) and core of the waveguide (green curve). The colored regions correspond to the normalized absorption spectra obtained when illuminating the structure from the substrate with a plane wave and mapped into $\beta = k_0 n_{sub} \sin\theta_{inc}\Lambda/\pi$ (see reference [28] for a detailed explanation). The blue triangles curve in the guided region correspond to the fundamental TM_0 photonic mode of the isolated waveguide. The array of MNW supports plasmonic chain modes. The first one is a

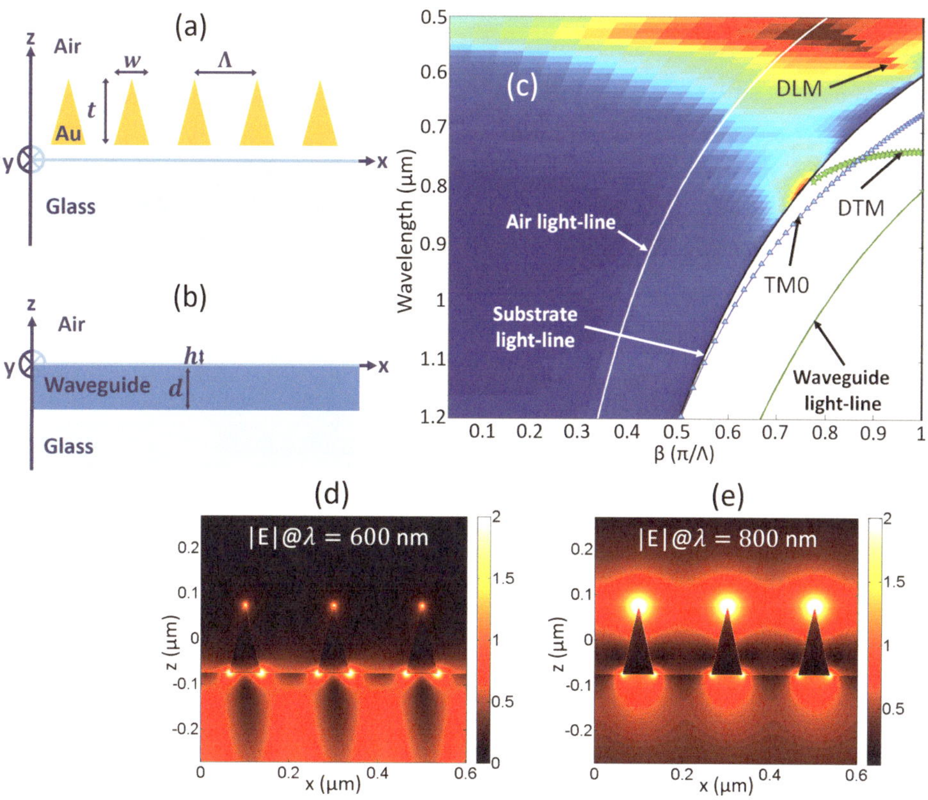

Figure 13.
Schematic representation of (a) periodic array of gold nanowires with triangular cross section ($w = 72$ nm, $t = 144$ nm, $\Lambda = 200$ nm) on top of a glass substrate surrounded by air, and (b) a dielectric waveguide of thickness $d = 200$ nm and refractive index $n_{wg} = 2.0$ buried a depth $h_1 = 30$ nm in a glass substrate. (c) Dispersion curves of the TM_0 mode supported by the waveguide (blue triangles), dipolar longitudinal mode (radiated to the substrate) and dipolar longitudinal mode (green stars). Energy density maps of (d) DLM at $\lambda = 600$ nm and (e) DTM at $\lambda = 800$ nm.

dipolar longitudinal mode (DLM) radiating into the substrate (which is barely excited at the Bragg condition). This mode is characterized by charges coupling and electromagnetic field enhancement at the base (bottom apexes) of the nanowires, as can be observed in the energy density map in **Figure 13d**. The second plasmonic chain mode (green stars) is a dipolar transverse mode (DTM) resulting from coupling of dipoles oriented along the z axis. This guided mode is characterized by a strong electromagnetic field enhancement at the upper apexes of the nanowires, as can be observed in **Figure 13e**. We can notice in the dispersion curves that at $\lambda =$ 747 nm ($\beta = 0.86\pi/\Lambda$), the DTM crosses the TM_0 mode of the waveguide, situation that will lead to a strong coupling between these modes when integrating both waveguiding systems.

When integrating the MNW on top of the dielectric waveguide (**Figure 14a**), the dispersion curves shows three modes in the guided region (**Figure 14b**), one corresponding to the DLM (magenta circles) and two other branches corresponding to the anti-symmetric (blue circles) and symmetric (red dots) dipolar transverse chain modes. The mode splitting between these two last resonances arises from strong coupling between dipolar transverse and TM_0 modes. The near field maps in **Figure 14c** present the amplitude and phase of the H_y field for the DLM, where is observed a field enhancement below each nanowire and in the waveguide region. For the antisymmetric DTM (**Figure 14d**) the field is mainly enhanced in the waveguide region, with a phase difference of π rad between the MNW and the waveguide; for the symmetric DTM (**Figure 14e**), the field is enhanced between the

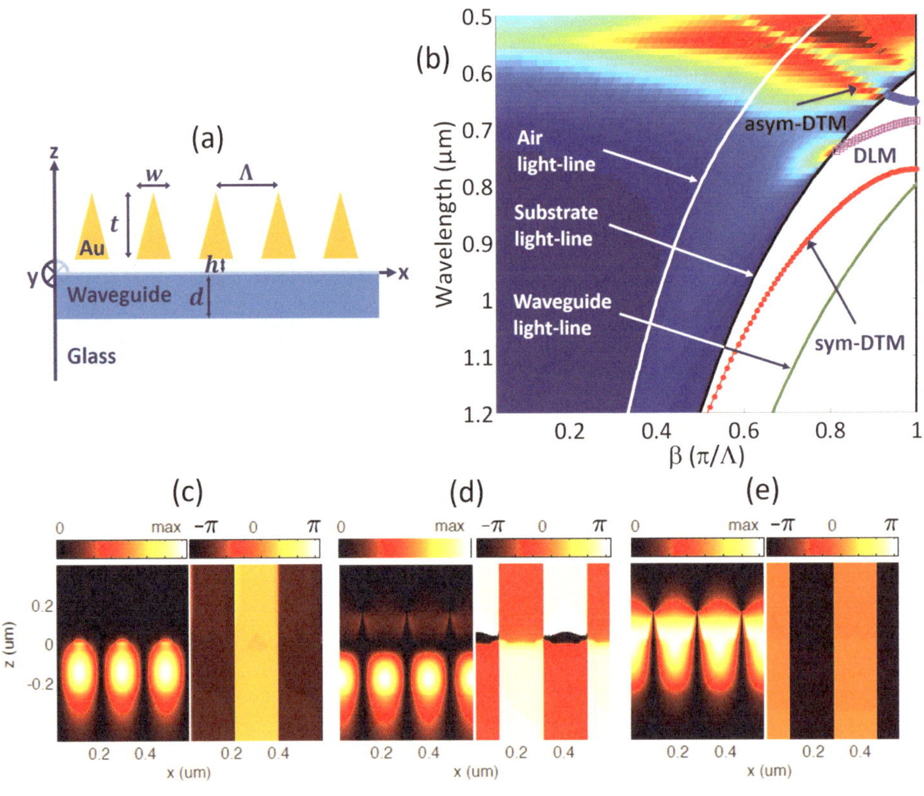

Figure 14.
(a) Schematic representation of an infinite periodic array of MNW integrated on top of a dielectric waveguide buried in a glass substrate. The superstrate is air. (b) In the dispersion curves are observed three chain modes in the guided region: DLM (magenta circles), asymmetric DTM (blue circles) and symmetric DTM (red dots). Distribution of the amplitude and phase of magnetic field in the out of plane direction of the (c) DLM, (d) asymmetric DTM and (e) symmetric DTM.

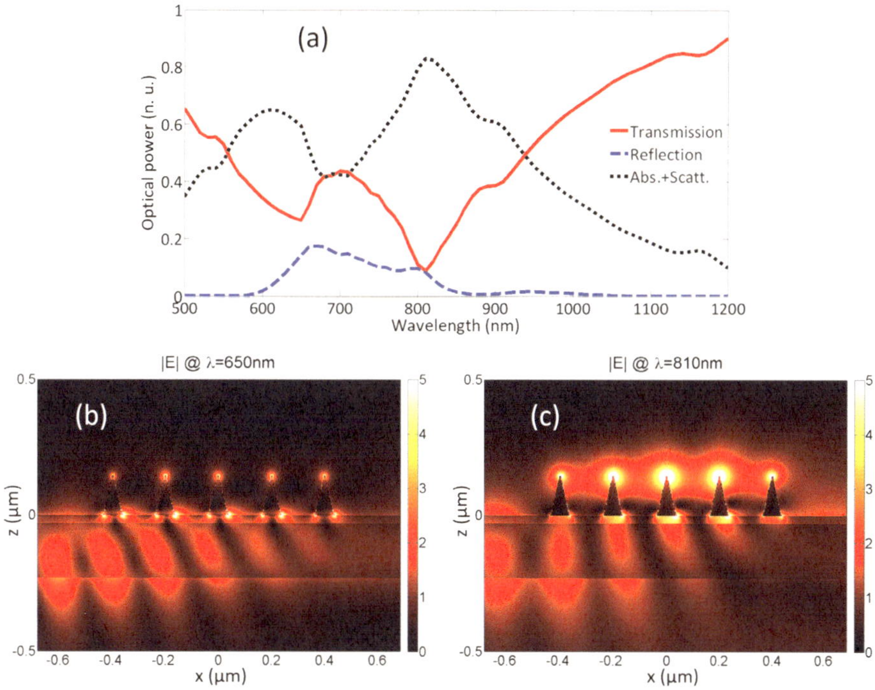

Figure 15.
(a) Normalized transmission (red), reflection (blue dashed) and absorption plus scattering (black dotted) curves of a periodic array of 5 gold nanowires integrated on top of the dielectric waveguide. (b) At $\lambda = 650\,nm$, the energy density map shows a field enhancement at the lower apexes of the MNW. (c) At $\lambda = 810\,nm$, the energy density map shows a strong enhancement of the field at the upper apexes of the MNW.

MNW with no phase difference between waveguide and MNW, being defined the symmetry of the modes.

To corroborate the coupling between the photonic and plasmonic chain modes, we simulated the beam propagation along the integrated structure. For this simulation, we used a finite number of 5 MNW and excited the waveguide with its fundamental TM_0 mode. The transmission (red curve), reflection (blue dashed curve) and absorption plus scattering (black dotted curve) where normalized to the incident electromagnetic field and the results are shown on **Figure 15a**.

The transmission shows a first minimum value around $\lambda = 650$ nm, corresponding to the excitation of DLM and antisymmetric DTM. This excitation if corroborated in the energy density map of **Figure 15b**, where is observed a field enhancement at the bottom apexes of the nanowires. The second minimum observed in the transmission spectrum around $\lambda = 810$ nm, correspond to the efficient excitation of the symmetric DTM, being characterized by a strong enhancement of the electromagnetic field at the upper apexes of the nanowires, as can be observed in the energy density map of **Figure 15c**. According to the numerical calculations, at this wavelength the amplitude of the electromagnetic field measured 10 nm above the apex of the nanowires, is 8.7 times stronger in comparison to the amplitude of the incident electromagnetic field. This field enhancement is referred as tip-localized surface plasmon resonance.

4. Conclusions

As we have studied in this chapter, mode hybridization between plasmonic and photonic guided modes offers the possibility to design integrated devices for light confinement in nanometric volumes. Also, as light propagates in dielectric structures, these integrated devices reduce intrinsic propagation losses in conventional plasmonic waveguides.

When properly excited, we have studied how localized surface plasmons can couple in periodic arrays of metallic nanowires, leading to light propagation. In other words, periodic arrays of MNW can behave as plasmonic waveguides. Depending on the geometry of their cross section (shape and aspect ratio), the electromagnetic field can be strongly confined and localized, desired property for number of applications, like excitation of quantum dots or single photon emitters, surface enhanced Raman spectroscopy, and biosensing.

The examples and explanations brought in this chapter can also be expanded to other geometries and material combinations. The reader should always remind that mode hybridization is the core of the physics behind the design of integrated photonic-plasmonic devices for light guiding applications.

These hybrid photonic-plasmonic systems offers the capability of matching diffraction limited guided optics, with nanometric materials, opening new perspectives for the development of a new generation of integrated optical devices.

Acknowledgements

The authors thank National Council of Science and Technology, CONACYT, for partial financial support (Basic Scientific Research, Grant No. A1-S-21527).

Conflict of interest

The authors declare no conflicts of interest.

Author details

Ricardo Téllez-Limón[1]* and Rafael Salas-Montiel[2]

1 CONACYT – Center for Scientific Research and Higher Education at Ensenada (CICESE), Unit Monterrey, Nuevo Leon, Mexico

2 L2n – Laboratory Light, Nanomaterials and Nanotechnologies, CNRS ERL 7004 and University of Technology at Troyes, Troyes, France

*Address all correspondence to: rtellez@conacyt.mx; rafael.salas@utt.fr

References

[1] Novotny L, Hecht B. Principles of Nano-Optics. 2nd ed. Cambridge: Cambridge University Press; 2012. 564 p. DOI: 10.1017/CBO9780511794193

[2] Lévêque G, Martin, O. J. F. Optical interactions in a plasmonic particle coupled to a metallic film, Optics Express, 2006; 14:9971–9981. DOI: 10.1364/OE.14.009971

[3] Dong, J-W, Deng, Z-L. Direct eigenmode analysis of plasmonic modes in metal nanoparticle chain with layered medium, Optics Letters, 2013; 38:2244–2246. DOI: 10.1364/OL.38.002244

[4] Quinten M, Leitner A, Krenn J. R, Aussenegg F. R. Electromagnetic energy transport via linear chains of silver nanoparticles, Optics Letters, 1998; 23: 1331–1333. DOI: 10.1364/OL.23.001331

[5] Brongersma M. L, Hartman J. W, Atwater H. A. Electromagnetic energy transfer and switching in nanoparticle chain arrays below the diffraction limit, Physical Review B, 2000; 62: R16356-R16359. DOI: 10.1103/PhysRevB.62.R16356

[6] Fung K. H, Chan C. T. Plasmonic modes in periodic metal nanoparticle chains: a direct dynamic eigenmode analysis, Optics Letters, 2007; 32:973–975. DOI: 10.1364/OL.32.000973

[7] Compaijen P. J, Malyshev V. A, Knoester J. Elliptically polarized modes for the unidirectional excitation of surface plasmon polaritons, Optics Express, 2016; 24:3858–3872. DOI: 10.1364/OE.24.003858

[8] Février M, Gogol P, Aassime A, Mégy R, Delacour C, Chelnokov A, Apuzzo A, Blaize S, Lourtioz, J-M, Dagens B. Giant Coupling Effect between Metal Nanoparticle Chain and Optical Waveguide, Nano Letters, 2012; 12: 1032–1037. DOI: 10.1021/nl204265f

[9] Saha S, Dutta A, Kinsey N, Kildishev A. V, Shalaev V. M, Boltasseva A. On-Chip Hybrid Photonic-Plasmonic Waveguides with Ultrathin Titanium Nitride Films, ACS Photonics, 2018; 5:4423–4431. DOI: 10.1021/acsphotonics.8b00885

[10] Saleh B. E. A, Teich M. C. Fundamentals of Photonics. 2nd ed. New York: John Wiley & Sons, Inc; 2007. 947 p. DOI: 10.1002/0471213748

[11] Maier S. A, Plasmonics: Fundamentals and Applications. New York: Springer US; 2007. 224 p. DOI: 10.1007/0–387–37825-1

[12] Jackson J. D. Classical electrodynamics. 3rd ed. New York: Wiley; 1998. 832 p. ISBN: 9780471309321

[13] Beltran Madrigal J, Tellez-Limon R, Gardillou F, Barbier D, Geng W, Couteau C, Salas-Montiel R, Blaize S. Hybrid integrated optical waveguides in glass for enhanced visible photoluminescence of nanoemitters, Applied Optics, 2016; 55:10263–10268. DOI: 10.1364/AO.55.010263

[14] Tellez-Limon R, Blaize S, Gardillou F, Coello V, Salas-Montiel R. Excitation of surface plasmon polaritons in a gold nanoslab on ion-exchanged waveguide technology, Applied Optics, 2020; 59:572–578. DOI: 10.1364/AO.381915

[15] Inclán Ladino A, Mendoza-Hernández J, Arroyo-Carrasco M. L, Salas-Montiel R, García-Méndez M, Coello V, Tellez-Limon R. Large depth of focus plasmonic metalenses based on Fresnel biprism, AIP Advances, 2020; 10: 045025. DOI: 10.1063/5.0004208

[16] Katsidis C. C, Siapkas D. I. General transfer-matrix method for optical multilayer systems with coherent,

partially coherent, and incoherent interference, Applied Optics, 2002; 41: 3978–3987. DOI: 10.1364/AO.41.003978

[17] Anemogiannis E, Glytsis E. N. Multilayer waveguides: efficient numerical analysis of general structures, Journal of Lightwave Technology, 1992; 10:1344–1351. DOI: 10.1109/50.166774

[18] Anemogiannis E, Glytsis E. N, Gaylord T. K. Determination of guided and leaky modes in lossless and lossy planar multilayer optical waveguides: reflection pole method and wavevector density method, Journal of Lightwave Technology, 1999; 17:929–941. DOI: 10.1109/50.762914

[19] Kocabaş Ş. E, Veronis G, Miller D. A. B, Fan S. Modal analysis and coupling in metal-insulator-metal waveguides, Physical Review B, 2009; 79: 035120. DOI: 10.1103/PhysRevB.79.035120

[20] Davis T. J. Surface plasmon modes in multi-layer thin-films, Optics Communications, 2009; 282:135–140. DOI: 10.1016/j.optcom.2008.09.043

[21] Nesterov M. L, Kats A. V, Turitsyn S. K. Extremely short-length surface plasmon resonance devices, Optics Express, 2008; 16: 20227–20240. DOI: 10.1364/OE.16.020227

[22] Kekatpure R. D, Hryciw A. C, Barnard E. S, Brongersma M. L. Solving dielectric and plasmonic waveguide dispersion relations on a pocket calculator, Optics Express, 2009; 17: 24112–24129. DOI: 10.1364/OE.17.024112

[23] Haeseler F. V, Peitgen H. O. Newton's method and complex dynamical systems, Acta Applicandae Mathematica, 1988; 13:3–58. DOI: 10.1007/BF00047501

[24] Garrido Alzar C. L, Martinez M. A. G, Nussenzveig P. Classical analog of electromagnetically induced transparency, American Journal of Physics; 2001; 70:37–41. DOI: 10.1119/1.1412644

[25] Novotny L. Strong coupling, energy splitting, and level crossings: A classical perspective, American Journal of Physics, 2010; 78:1199–1202. DOI: 10.1119/1.3471177

[26] Tellez-Limon R, Fevrier M, Apuzzo A, Salas-Montiel R, Blaize S. Modal analysis of LSP propagation in an integrated chain of gold nanowires. In: Proceedings SPIE Nanophotonic Materials X, 13 September 2013; San Diego. California: Proc. SPIE; 2013. p. 88070J

[27] Tellez-Limon R, Fevrier M, Apuzzo A, Salas-Montiel R, Blaize S. Theoretical analysis of Bloch mode propagation in an integrated chain of gold nanowires, Photonics Research, 2014; 2:24–30. DOI: 10.1364/PRJ.2.000024

[28] Tellez-Limon R, Février M, Apuzzo A, Salas-Montiel R, Blaize S. Journal of the Optical Society of America B, 2017; 34: 2147–2154. DOI: 10.1364/JOSAB.34.002147

[29] Hocker G. B, Burns W. K. Mode dispersion in diffused channel waveguides by the effective index method, Applied Optics, 1977; 16:113–118. DOI: 10.1364/AO.16.000113

[30] Payne F. P. A new theory of rectangular optical waveguides, Optical and Quantum Electronics, 1982; 14:525–537. DOI: 10.1007/BF00610308

[31] Patchett S, Khorasaninejad M, Nixon O, Saini S. S. Effective index approximation for ordered silicon nanowire arrays, Journal of the Optical Society of America B, 2013; 30:306–313. DOI: 10.1364/JOSAB.30.000306

[32] Szafranek D, Leviatan Y. A Source-Model Technique for analysis of wave guiding along chains of metallic

nanowires in layered media, Optics Express, 2011; 19:25397–25411. DOI: 10.1364/OE.19.025397

[33] Moharam M. G, Gaylord, T. K. Rigorous coupled-wave analysis of planar-grating diffraction, Journal of the Optical Society of America, 1981; 71: 811–818. DOI: 10.1364/JOSA.71.000811

[34] Moharam M. G, Grann E. B, Pommet D. A, Gaylord T. K. Formulation for stable and efficient implementation of the rigorous coupled-wave analysis of binary gratings, Journal of the Optical Society of America A, 1995; 12: 1068–1076. DOI: 10.1364/JOSAA.12.001068

[35] Liu H, Lalanne P. Comprehensive microscopic model of the extraordinary optical transmission, Journal of the Optical Society of America A, 2010; 27: 2542–2550. DOI: 10.1364/JOSAA.27.002542

[36] Noponen E, Turunen J. Eigenmode method for electromagnetic synthesis of diffractive elements with three-dimensional profiles, Journal of the Optical Society of America A, 1994; 11: 2494–2502. DOI: 10.1364/JOSAA.11.002494

[37] Li L. New formulation of the Fourier modal method for crossed surface-relief gratings, Journal of the Optical Society of America A, 1997; 14: 2758–2767. DOI: 10.1364/JOSAA.14.002758

[38] Kim H, Park J, Lee B. Fourier modal method and its applications in computational nanophotonics. 1st ed. Boca Raton: CRC Press; 2012. 326 p. DOI: 10.1201/b11710

[39] Bykov D. A, Bezus E. A, Doskolovich L. L. Use of aperiodic Fourier modal method for calculating complex-frequency eigenmodes of long-period photonic crystal slabs, Optics Express, 2017; 25: 27298–27309. DOI: 10.1364/OE.25.027298

Chapter 9

Gate-All-Around FETs: Nanowire and Nanosheet Structure

Jun-Sik Yoon, Jinsu Jeong, Seunghwan Lee, Junjong Lee and Rock-Hyun Baek

Abstract

DC/AC performances of 3-nm-node gate-all-around (GAA) FETs having different widths and the number of channels (N_{ch}) from 1 to 5 were investigated thoroughly using fully-calibrated TCAD. There are two types of GAAFETs: nanowire (NW) FETs having the same width (W_{NW}) and thickness of the channels, and nanosheet (NS) FETs having wide width (W_{NS}) but the fixed thickness of the channels as 5 nm. Compared to FinFETs, GAAFETs can maintain good short channel characteristics as the W_{NW} is smaller than 9 nm but irrespective of the W_{NS}. DC performances of the GAAFETs improve as the N_{ch} increases but at decreasing rate because of the parasitic resistances at the source/drain epi. On the other hand, gate capacitances of the GAAFETs increase constantly as the N_{ch} increases. Therefore, the GAAFETs have minimum RC delay at the N_{ch} near 3. For low power applications, NWFETs outperform FinFETs and NSFETs due to their excellent short channel characteristics by 2-D structural confinement. For standard and high performance applications, NSFETs outperform FinFETs and NWFETs by showing superior DC performances arising from larger effective widths per footprint. Overall, GAAFETs are great candidates to substitute FinFETs in the 3-nm technology node for all the applications.

Keywords: gate-all-around, nanowire, nanosheet, field-effect transistors, fin, RC delay, parasitic resistance, parasitic capacitance

1. Introduction

Gate-all-around (GAA) is a widely-using structure such as logic field-effect transistor (FET) due to its excellent short channel characteristics [1–6] or its high surface-to-volume ratio [7, 8], 3-D NAND flash memory for bit-cost scalability [9, 10], photodiode due to its waveguide effect [11, 12], and gas sensor due to its high physical fill factor or surface-to-volume ratio [13, 14]. Especially for logic applications, GAAFETs have been introduced by attaining good gate electronics and increasing current drivability under the same active area.

Currently, fin-shaped FETs (FinFETs) have been scaled down to 10-nm node [15] and further to 5-nm node [16] by forming ultra-sharp fin for high current drivability while maintaining gate-to-channel controllability. GAAFETs are possibly showing great potential to substitute FinFETs in the following technology node, and the performance comparisons between FinFETs and GAAFETs have been investigated [3–6, 17]. But more detailed analysis between FinFETs and GAAFETs is needed to set the device guideline by considering fine TCAD calibration and middle-of-line levels.

Therefore, in this work, DC/AC performances of 3-nm-node GAAFETs were investigated using fully-calibrated TCAD platform. By changing the GAA geometries, we found optimal GAA structure to minimize the RC delay for three different applications such as low power (LP), standard performance (SP), and high performance (HP) applications.

2. Device structure and simulation methods

All the simulation works were performed using Sentaurus TCAD [18]. Drift diffusion transport equations were calculated self-consistently with Poisson and electron/hole continuity equations. Density-gradient model was adopted for the quantum confinement of carriers within the channel. Slotboom bandgap narrowing model was used to consider the doping-dependent energy bandgap. Mobility models include Lombardi for the mobility degradation at the channel/oxide interface, inversion and accumulation layer model for impurity, phonon, and surface roughness scatterings, and low-field ballistic model for quasi-ballistic effects in ultra-short gate length (L_g). Shockley-Read-Hall, Auger, and Hurkx band-to-band tunneling recombination models were adopted. Deformation potential model was used to consider the stress-induced energy bandgap, effective mass, and effective density-of-states. All these physical models were used equivalently in [19, 20].

Figure 1 shows the schematic diagrams of FinFETs and three-stacked GAAFETs. FinFETs have highly-doped punch-through-stopper (PTS) at 2×10^{18} and 4×10^{18} cm^{-3} for NFETs and PFETs, respectively, in order to prevent the sub-fin leakage currents at off state [21, 22]. GAAFETs, on the other hand, have buried oxide (BOX) layer beneath the source/drain (S/D) regions without PTS so that the bottom leakage currents are completely blocked [1, 23]. Bulk FinFETs can adopt the BOX layer according to [24], but the conventional device structure

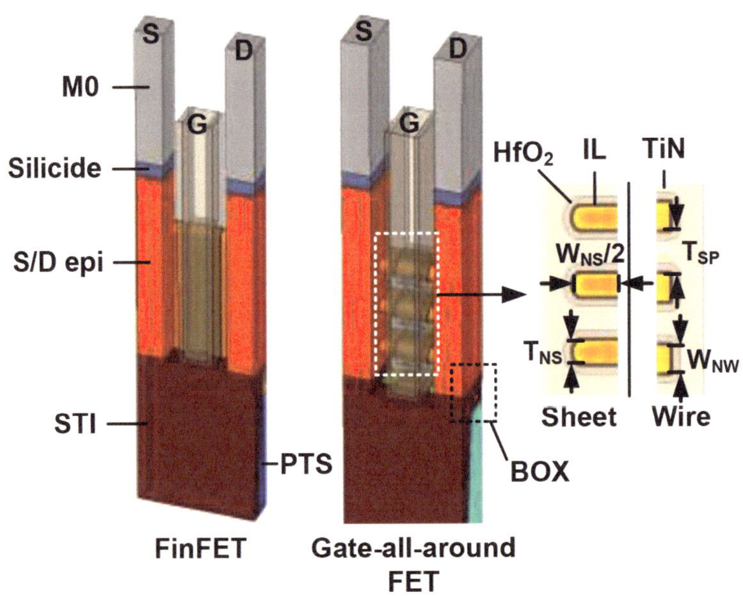

Figure 1.
Schematic diagrams of FinFETs and GAAFETs. 2-D cross-sections of nanosheet and nanowire channels were also specified to the right.

was considered in this work. S/D doping concentrations of the n-type and p-type devices are 2×10^{20} and 4×10^{20} cm^{-3}, respectively. Interfacial layer (IL), HfO_2, and low-k spacer regions have the dielectric constants of 3.9, 22.0, and 5.0, respectively. Contact resistivity at S/D and silicide interface is fixed to 10^{-9} $\Omega \cdot cm^2$ [25]. Equivalent oxide thickness (EOT) is 1.0 nm, which consists of 0.7-nm-thick IL and 1.7-nm-thick HfO_2.

Table 1 shows the geometrical parameters and values of 3-nm-node FinFETs and GAAFETs. Contacted poly pitch (CPP) and fin pitch (FP) are 42 and 21 nm, following 3-nm technology node [5]. There are two types of GAAFETs: nanowire FETs (NWFETs) having the same width and thickness as W_{NW}, and nanosheet FETs (NSFETs) having thin NS thickness (T_{NS}) of 5 nm but wide NS width (W_{NS}) as 10, 20, 30, 40, and 50 nm. The number of NW or NS channels (N_{ch}) is varied as 1, 2, 3, 4, and 5.

Figure 2 shows the schematic process flows of GAAFETs. The detailed gate-las process flows are described in [1]. After depositing $Si_{0.7}Ge_{0.3}$/Si multi-layer and etching like fin structure, poly-Si gate and low-k regions are formed. Inner-spacer is formed by etching sidewalls of $Si_{0.7}Ge_{0.3}$ regions selectively and depositing low-k regions. Followed by depositing BOX layer, selective epitaxial growth of S/D regions is performed. After removing poly-Si gate, channel release process is performed by etching $Si_{0.7}Ge_{0.3}$ regions selectively. Replacement metal gate, silicidation, and metal contact formations are done afterwards.

All the TCAD results were calibrated to Intel 10-nm node FinFETs [15]. Detailed calibration flows are as follows. Geometrical parameters such as L_g, fin width (W_{fin}), fin height (H_{fin}), CPP, and FP were referred from [15]. Subthreshold characteristics such as subthreshold swing (SS) and drain-induced barrier lowering (DIBL) were fitted by changing annealing temperature and time for proper S/D doping profiles. Saturation velocity was tuned to fit the drain current (I_{ds}) in the saturation region, whereas minimum low-field mobility and ballistic coefficient were varied to fit the I_{ds} in the linear region. Some parameters related to surface roughness scatterings were also modified to fit the I_{ds} in the strong inversion region accordingly. These calibration flows were equivalent as in [26]. After calibration, FinFETs were scaled down to the 3-nm node for comparison with GAAFETs.

Geometrical parameters		Values
CPP	Contacted poly pitch	42 nm
FP	Fin pitch	21 nm
NP	Nanowire/sheet pitch	W_{NW} or W_{NS} + 16 nm
L_g	Gate length	12 nm
L_{sp}	Spacer length	5 nm
W_{fin}	Fin width	5 nm
H_{fin}	Fin height	46 nm
W_{NW}	Nanowire width	5, 6, 7, 8, 9, 10 nm
W_{NS}	Nanosheet width	10, 20, 30, 40, 50 nm
T_{NS}	Nanosheet thickness	5 nm
T_{SP}	Nanowire/sheet spacing	10 nm
N_{ch}	The number of channels	1, 2, 3, 4, 5

Table 1.
Geometrical parameters and values of FinFETs and GAAFETs.

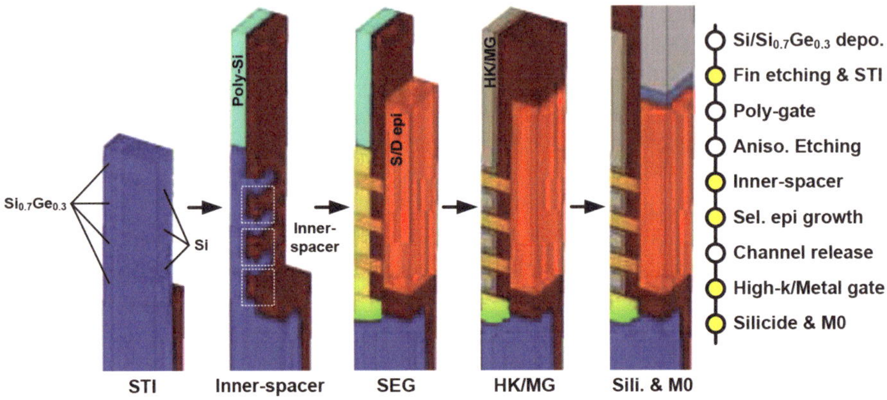

Figure 2.
Process flows of GAAFETs. Key process schemes of GAAFETs are $Si_{0.7}Ge_{0.3}$/Si multi-layer stacking, inner-spacer formation, and channel release by etching $Si_{0.7}Ge_{0.3}$ regions selectively.

3. Results and discussion

3.1 DC performances of NWFETs and NSFETs

Figure 3 shows the I_{ds} of all the GAAFETs having different W_{NW} or W_{NS} at the fixed N_{ch} of 3 at the drain voltages (V_{ds}) of 0.70 V. It is not shown in this figure, but the I_{ds} increases generally as the W_{NW} or W_{NS} increases irrespective of N_{ch}. As the W_{NW} increases, the I_{ds} shifts leftward and the gate-induced drain leakage (GIDL) increases by losing the gate-to-channel controllability [27]. P-type NWFETs have larger GIDL than n-type NWFETs due to larger S/D doping penetrations into the channel for p-type devices. On the other hand, NSFETs have small GIDL and I_{ds} shifts as thin T_{NS} of 5 nm forms 1-D structural confinement and maintains good short channel characteristics. To the following, there are three applications at different off-state currents (I_{off}): LP at the I_{off} of 100 pA/μm, SP at the I_{off} of 10 nA/μm, and HP at the I_{off} of 100 nA/μm [28]. These values were normalized to NP.

Figure 4 shows SS and DIBL of all the devices. Threshold voltages (V_{th}) and SS are extracted at the constant current of $W_{eff}/L_g \times 10^8$ A, where W_{eff} is the effective width equal to $2 \times H_{fin} + W_{fin}$ for FinFETs, $4 \times W_{NW} \times N_{ch}$ for NWFETs, and $(2 \times W_{NS} + 2 \times T_{NS}) \times N_{ch}$ for NSFETs. DIBL is calculated as the difference of the V_{th} at two different V_{ds} of 0.05 and 0.70 V for n-type (−0.05 and − 0.70 V for p-type) devices [29]. NWFETs degrade the short channel characteristics much than FinFETs as the W_{NW} is 9 and 10 nm. NSFETs, on the other hand, have smaller SS and DIBL than FinFETs even as the W_{NS} increases up to 50 nm because the gate-to-channel controllability is maintained by GAA structure and thin T_{NS} of 5 nm. But when the NWFETs have ultra-small W_{NS} of 5 or 6 nm, 2-D structural confinement decreases the SS and DIBL greatly, which would be preferable for LP applications. It is not shown in this figure, but the SS and DIBL are independent of N_{ch}.

Figure 5 summarizes the effective currents (I_{eff}) of n-type (top) and p-type (bottom) GAAFETs having different W_{NW} (or W_{NS}) and N_{ch}. I_{eff} was calculated using two I_{ds} at different V_{ds} and gate voltages (V_{gs}) as

$$I_{eff} = (I_H - I_L) / \ln\left(\frac{I_H}{I_L}\right) \quad (1)$$

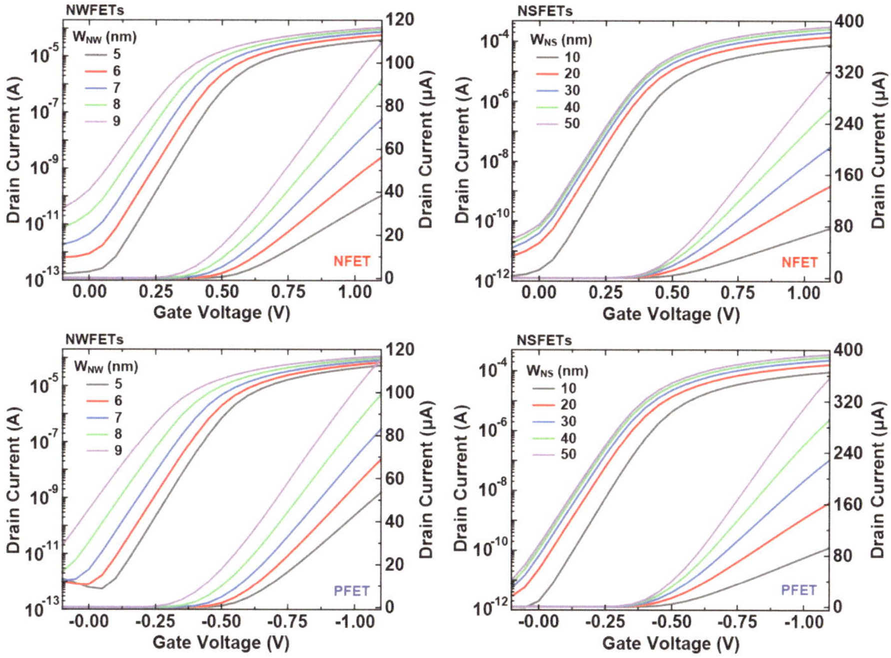

Figure 3.
I_{ds} of n-type (top) and p-type (bottom) NWFETs and NSFETs having different W_{NW} or W_{NS} at the fixed N_{ch} of 3 at the drain voltages (V_{ds}) of 0.70 V. it is not shown in this figure, but the GAAFETs have the same I_{ds} trends irrespective of N_{ch} (I_{ds} increases as the W_{NW} or W_{NS} increases).

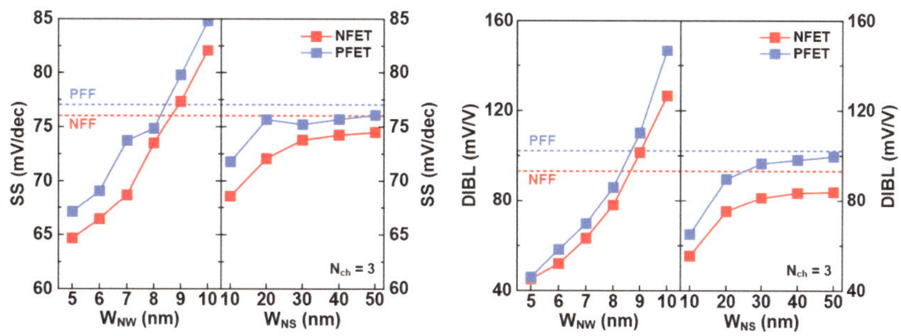

Figure 4.
SS (left) and DIBL (right) of FinFETs, NWFETs, and NSFETs having fixed N_{ch} of 3. It is not shown in this figure, but the GAAFETs have the same SS and DIBL irrespective of N_{ch}.

where $I_H = I_{ds}$ ($V_{gs} = V_{DD}$, $V_{ds} = V_{DD}/2$) and $I_L = I_{ds}$ ($V_{gs} = V_{DD}/2$, $V_{ds} = V_{DD}$) [30], and V_{DD} is the operation voltage fixed to 0.7 V. All the I_{eff} were normalized to the NP, and the I_{off} were fixed to 10 nA/μm for SP applications. GAAFETs need to have at least the N_{ch} of 3 to outperform the FinFETs. As the W_{NW} is 9 nm, both n-type and p-type NWFETs suffer from short channel effects (SCEs) and thus have smaller I_{eff} than the devices having smaller W_{NW} in spite of larger W_{eff}. NSFETs, on the other hand, have larger I_{eff} as the W_{NS} is larger as the SCEs are reduced by thin T_{NS} of 5 nm. But even though small same SS and DIBL are maintained for all the N_{ch}, the increasing rate of I_{eff} as a function of N_{ch} decreases as N_{ch} increases.

Figure 6 shows the S/D parasitic resistance (R_{sd}) of the GAAFETs having the W_{NW} or 7 nm and the W_{NS} of 30 nm as a function of N_{ch}. Other W_{NW} and W_{NS} have the same R_{sd} trends and thus are not shown in this work. R_{sd} was possibly extracted using Y-function method due to the linearity of Y-function at high V_{gs} [31]. As the

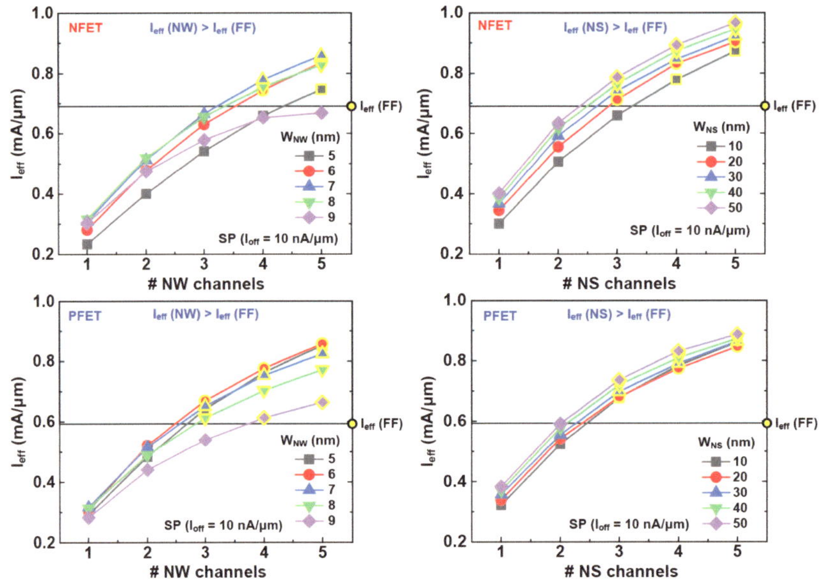

Figure 5.
I_{eff} of n-type (top) and p-type (bottom) GAAFETs having different W_{NW} (or W_{NS}) and N_{ch}. I_{eff} of n-type and p-type FinFETs are also specified as yellow symbols. Blue regions indicate that the GAAFETs have superior I_{eff} than the FinFETs.

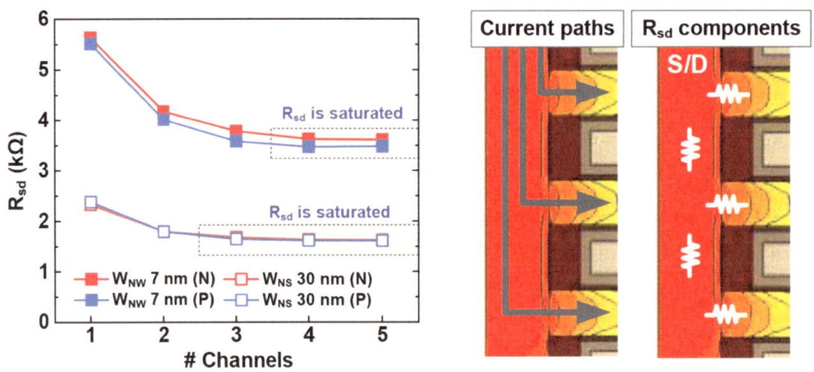

Figure 6.
R_{sd} of n-type and p-type GAAFETs having the W_{NW} of 7 nm and the W_{NS} of 30 nm as a function of N_{ch} (left) and the 2-D schematic diagram of half of the GAAFETs showing the current paths and R_{sd} components (right).

N_{ch} increases, R_{sd} of the GAAFETs decrease but at decreasing rate. Furthermore, R_{sd} becomes saturated as the N_{ch} is 3 or 4. This phenomena can be explained by 2-D schematic diagrams shown in the right of **Figure 6**. Since the S/D contacts reside at the top of the S/D epi, current paths start from the top toward the channels at the bottom. As the N_{ch} increases, longer current paths are needed to flow the bottom-side channels, facing more R_{sd} components at the S/D epi. Thus, increasing the N_{ch} beyond 3 or 4 does not help DC performance improvements greatly.

3.2 AC performances of NWFETs and NSFETs

Figure 7 summarizes the gate capacitances (C_{gg}) of all the GAAFETs. The C_{gg} is extracted at the V_{gs} and the V_{ds} of V_{DD}. Generally, C_{gg} increases as the W_{NW} (or W_{NS}) or N_{ch} increases due to the increased W_{eff}. PFETs have larger C_{gg} than NFETs due to larger S/D doping concentrations and penetrations into the channels. Different

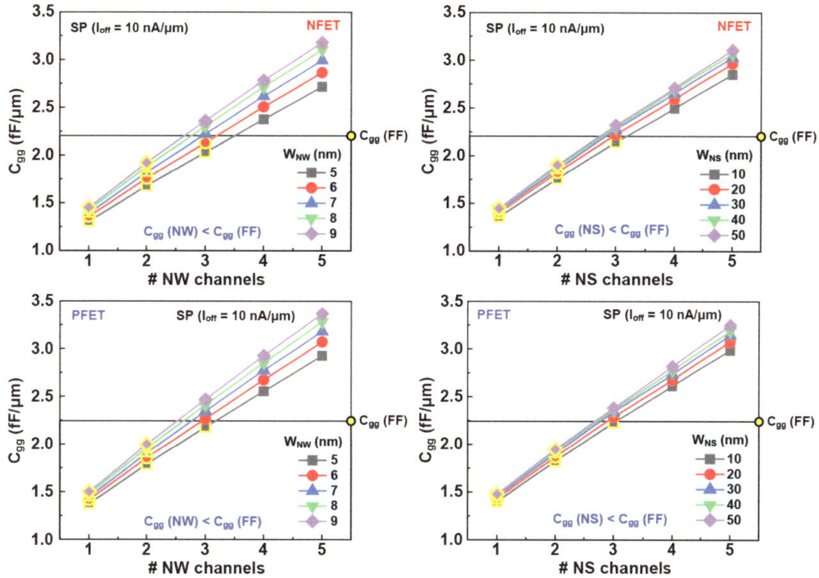

Figure 7.
C_{gg} of n-type (top) and p-type (bottom) GAAFETs having different W_{NW} (or W_{NS}) and N_{ch}. C_{gg} of n-type and p-type FinFETs are also specified as yellow symbols. Blue regions indicate that the GAAFETs have smaller C_{gg} than the FinFETs.

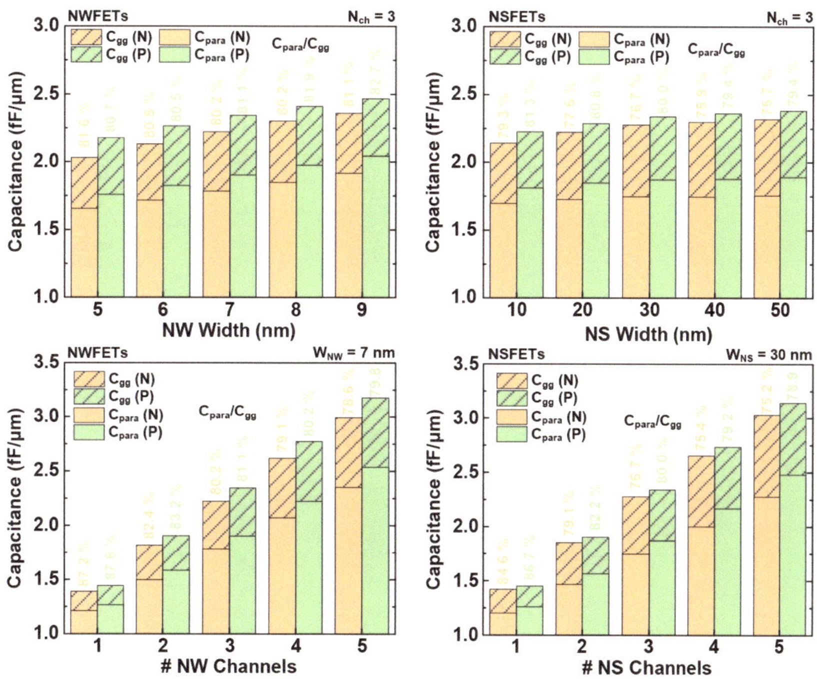

Figure 8.
C_{gg} and C_{para} of NWFETs (left) and NSFETs (right) having different W_{NW} (or W_{NS}) at the fixed N_{ch} of 3 and having different N_{ch} at the fixed W_{NW} of 7 nm (or W_{NS} of 30 nm). Percentages represent the C_{para}/C_{gg}.

from the I_{eff} trends, the GAAFETs have N_{ch} smaller than 3 to outperform the FinFETs, thus there are performance trade-offs between I_{eff} and C_{gg} as a function of N_{ch}. Furthermore, the increasing rate of C_{gg} as a function of N_{ch} is constant while the increasing rate of I_{eff} as a function of N_{ch} decreases, which would degrade the RC delay ($= I_{eff}V_{DD}/C_{gg}$) as the N_{ch} increases.

Figure 8 shows the C_{gg} and parasitic capacitances (C_{para}) of the GAAFETs varying N_{ch} and W_{NW} (or W_{NS}). C_{para} is extracted at off-state for SP applications. For all the cases, PFETs have larger C_{para} than NFETs due to larger S/D doping and penetrations into the channels [20]. At the fixed N_{ch} of 3, larger W_{NW} or W_{NS}, except for p-type NWFETs, decreases the C_{para}/C_{gg} because the proportion of the channels out of the metal gate increases. For the same reason, larger N_{ch} decreases the C_{para}/C_{gg}. Large C_{para}/C_{gg} at the W_{NW} of 9 nm for NFETs is because large SS forms on state before reaching strong inversion region.

Figure 9 shows the S/D doping profiles of NFETs (top) and PFETs (bottom) having different W_{NW} at the fixed N_{ch} of 3. In general, NFETs have larger doping concentrations in the middle of channels than PFETs because the Ge intermixing within multi-stacked $Si/Si_{0.7}Ge_{0.3}$ layers increases the Ge concentration at the channels and assists more phosphorus dopants diffusing into the channels while it segregates boron dopants [32–34]. Both NFETs and PFETs increase the doping concentrations in the middle of channels as the W_{NW} increases because the dopant segregations near the low-k spacer regions decrease [35]. But PFETs increase the doping concentrations in the middle of channels much due to smaller Ge intermixing for larger W_{NW}. This great increase of the doping concentrations in the middle of channels increases the C_{para}/C_{gg} for p-type NWFETs (as shown in **Figure 8**).

Figure 10 finalizes the RC delay of all the GAAFETs for LP, SP, and HP applications. N-type FinFETs have smaller RC delay than p-type FinFETs for all the applications due to better short channel characteristics, greater I_{eff} (as shown in **Figure 5**) and smaller C_{gg} (as shown in **Figure 8**). For LP applications, n-type GAAFETs having

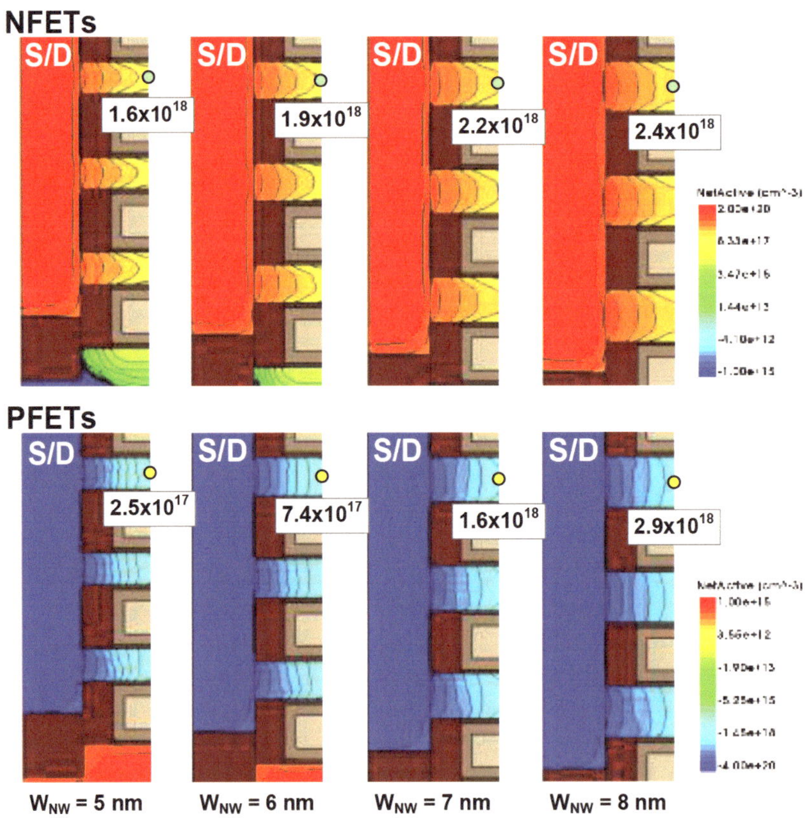

Figure 9.
S/D doping profiles of NFETs (top) and PFETs (bottom) having different W_{NW} at the fixed N_{ch} of 3. Doping concentrations in the middle of top-side channels are also specified.

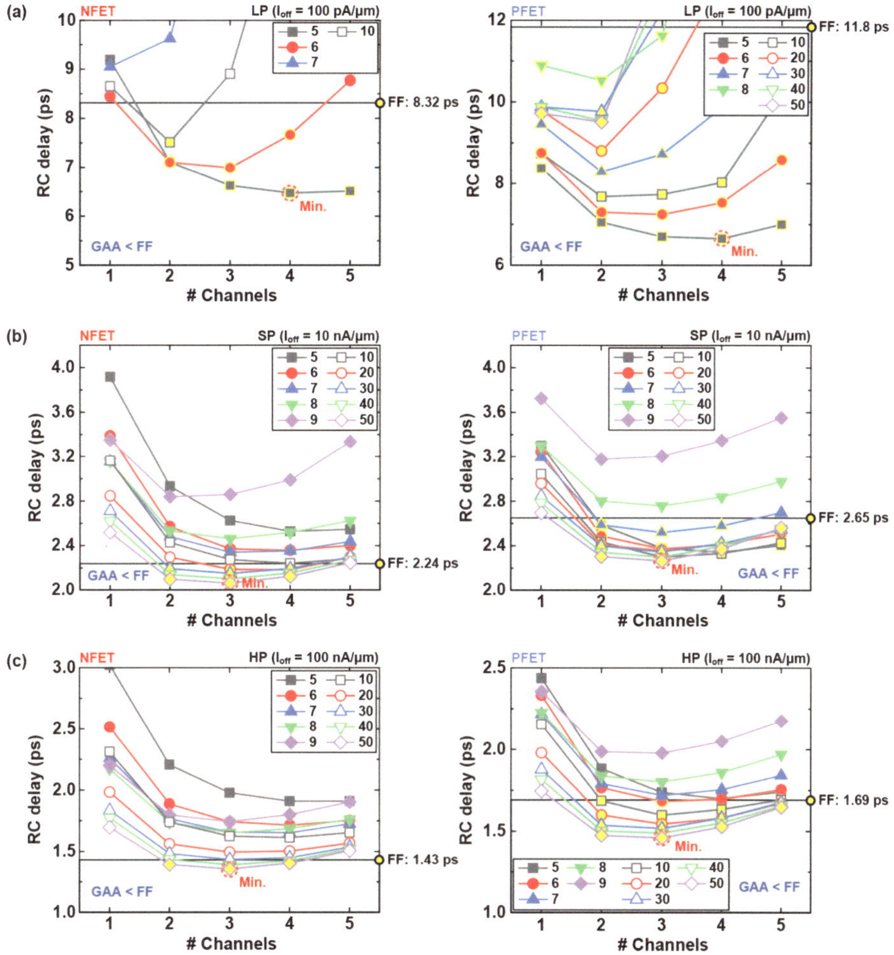

Figure 10.
RC delay of all the GAAFETs for (a) LP, (b) SP, and (c) HP applications. RC delay of FinFETs for three different applications are also specified. The devices having the RC delay smaller than FinFETs are marked as yellow.

small W_{NW} equal to 5 or 6 nm can outperform n-type FinFETs by decreasing SS and DIBL critically. But as the N_{ch} is 1 (or 5), the I_{eff} decreases greatly (or the C_{gg} increases greatly), thus degrading the RC delay. On the other hand, p-type GAAFETs have more W_{NW} or W_{NS} options to outperform p-type FinFETs because boron dopants of the GAAFETs are segregated by $Si/Si_{0.7}Ge_{0.3}$ intermixing and have more abrupt S/D doping profile than p-type FinFETs. For LP applications, both n- and p-type GAAFETs have the minimum RC delay at the W_{NW} of 5 nm and the N_{ch} of 4. For both SP and HP applications, both n- and p-type GAAFETs have the minimum RC delay at the W_{NS} of 50 nm and the N_{ch} of 3. As the W_{NS} increases beyond 50 nm, RC delay decrease but a little (as shown in **Appendix**). All these RC delay are achieved by enhancing the I_{eff} rather than the C_{gg}. To outperform the FinFETs, therefore, GAAFETs should be NWFETs, showing outstanding short channel characteristics, for LP applications and NSFETs, showing superior DC performance, for SP and HP applications.

4. Conclusion

3-nm-node GAAFETs have been analyzed by changing W_{NW} (or W_{NS}) and N_{ch} using fully-calibrated TCAD. Compared to FinFETs, GAAFETs have smaller and

SS and DIBL as the W_{NW} is smaller than 9 nm but irrespective of the W_{NS}. Both I_{eff} and C_{gg} of the GAAFETs increase as the N_{ch} increases, but the increasing rate of I_{eff} decreases due to the increase of R_{sd} at the longer S/D epi. The increasing rate of C_{gg}, on the other hand, is almost constant. Because of these phenomena, Minimum RC delay are formed at the middle N_{ch} of 3 or 4. The NWFETs having the W_{NW} of 5 or 6 nm achieve smaller RC delay than the FinFETs by achieving better gate electronics for LP applications, whereas the NSFETs having the W_{NS} of 40 or 50 nm increase the I_{eff} greatly and thus decrease the RC delay for SP and HP applications. Overall, GAAFETs are possible candidates to substitute FinFETs in the 3-nm technology node for all the applications by adopting different W_{NW} or W_{NS}.

Conflict of interest

The authors declare no conflict of interests.

Appendices and nomenclature

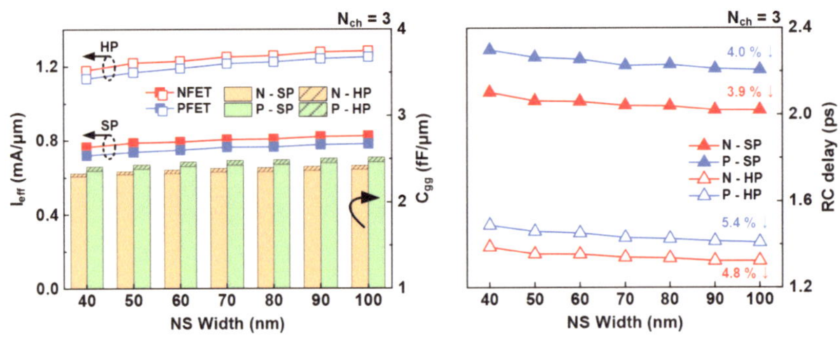

Figure A1.
I_{eff}, C_{gg}, and RC delay of the NSFETs having the W_{NS} of 40, 50, 60, 70, 80, 90, and 100 nm at the fixed N_{ch} of 3 for SP and HP applications.

Figure A1 shows the DC/AC performances of the NSFETs as the W_{NS} increases from 40 to 100 nm. Minimum RC delay are formed at the W_{NS} of 50 nm and the N_{ch} of 3 as shown in **Figure 10**, but much smaller RC delay can be attained as the W_{NS} increases to 100 nm by increasing the I_{eff} rather than the C_{gg} even though larger W_{NS} extends the device area. For the most, RC delay decrease by 5.4% for PFETs as the W_{NS} increases from 40 to 100 nm.

Author details

Jun-Sik Yoon, Jinsu Jeong, Seunghwan Lee, Junjong Lee and Rock-Hyun Baek*
Electrical Engineering, Pohang University of Science and Technology, Pohang, Republic of Korea

*Address all correspondence to: rh.baek@postech.ac.kr

References

[1] Loubet N, Hook T, Montanini P, Yeung C.-W, Kanakasabapathy S, Guillorn M, Yamashita T, Zhang J, Miao X, Wang J, Young A, Chao R, Kang M, Liu Z, Fan S, Hamieh B, Sieg S, Mignot Y, Xu W, Seo S.-C, Yoo J, Mochizuki S, Sankarapandian M, Kwon O, Carr A, Greene A, Park Y, Frougier J, Galatage R, Bao R, Shearer J, Conti R, Song H, Lee D, Kong D, Xu Y, Arceo A, Bi Z, Xu P, Muthinti R, Li J, Wong R, Brown D, Oldiges P, Wu T, Gupta D, Lian S, Divakaruni R, Gow T, Labelle C, Lee S, Paruchuri V, Bu H, Khare M. Stacked nanosheet gate-all-around transistor to enable scaling beyond FinFET. In: Proceedings of 2017 Symposium on VLSI Technology, Kyoto, 2017, pp. T230-T231, DOI: 10.23919/VLSIT.2017.7998183.

[2] Yoon J.-S, Rim T, Kim J, Meyyappan M, Baek C.-K, and Jeong Y.-H. Vertical gate-all-around junctionless nanowire transistors with asymmetric diameters and underlap lengths. Applied Physics Letters. 2014;105:102105-1-4. DOI: 10.1063/1.4895030.

[3] Lee Y. M, Na M. H, Chu A, Young A, Hook T, Liebmann L, Nowak E. J, Baek S. H, Sengupta R, Trombley H, and Miao X. Accurate performance evaluation for the horizontal nanosheet standard-cell design space beyond 7nm technology. In: Proceedings of 2017 IEEE International Electron Devices Meeting (IEDM), San Francisco, CA, 2017, pp. 29.3.1-29.3.4, DOI: 10.1109/IEDM.2017.8268474.

[4] Barraud S, Lapras V, Previtali B, Samson M. P, Lacord J, Martinie S, Jaud M.-A, Athanasiou S, Triozon F, Rozeau O, Hartmann J. M, Vizioz C, Comboroure C, Andrieu F, Barbé J. C, Vinet M. Ernst T. Performance and design considerations for gate-all-around stacked-nanowires FETs. In: Proceedings of 2017 IEEE International Electron Devices Meeting (IEDM), San Francisco, CA, 2017, pp. 29.2.1-29.2.4, DOI: 10.1109/IEDM.2017.8268473.

[5] Yakimets D, Garcia Bardon M, Jang D, Schuddinck P, Sherazi Y, Weckx P, Miyaguchi K, Parvais B, Raghavan P, Spessot A, Verkest D, Mocuta A. Power aware FinFET and later nanosheet FET targeting for 3nm CMOS technology. In: Proceedings of 2017 IEEE International Electron Devices Meeting (IEDM), San Francisco, CA, 2017, pp. 20.4.1-20.4.4, doi: 10.1109/IEDM.2017.8268429.

[6] Song S. C, Colombeau B, Bauer M, Moroz V, Lin X-W, Asenov P, Sherlekar D, Choi M, Huang J, Cheng B, Chidambaram C, Natarajan S. 2nm node: benchmarking FinFET vs nano-slab transistor architectures for artificial intelligence and next gen smart mobile devices. In: Proceedings of 2019 Symposium on VLSI Technology, Kyoto, Japan, 2019, pp. T206-T207. DOI: 10.23919/VLSIT.2019.8776478.

[7] Yoon J.-S, Kim K, Baek C.-K. Core-shell homojunction silicon vertical nanowire tunneling field-effect transistors. Scientific Reports. 2017;7:41142-1-9. DOI: 10.1038/srep41142.

[8] Yoon J.-S, Kim K, Meyyappan M, Baek C.-K. Bandgap engineering and strain effects of core-shell tunneling field-effect transistors. IEEE Transactions on Electron Devices. 2018;65:277-281. DOI: 10.1109/TED.2017.2767628.

[9] Tanaka H, Kido M, Yahashi K, Oomura M, Katsumata R, Kito M, Fukuzumi Y, Sato M, Nagata Y, Matsuoka Y, Iwata Y, Aochi H, Nitayama A. Bit cost scalable technology with punch and plug process for ultra high density flash memory.

In: Proceedings of 2007 IEEE Symposium on VLSI Technology, Kyoto, 2007, pp. 14-15. DOI: 10.1109/VLSIT.2007.4339708.

[10] Kim J, Hong A. J, Kim S. M, Song E. B, Park J. H, Han J, Choi S, Jang D, Moon J.-T, Wang K. L. Novel vertical-stacked-array-transistor (VSAT) for ultra-high-density and cost-effective NAND flash memory devices and SSD (solid state drive). In: Proceedings of 2009 Symposium on VLSI Technology, Honolulu, HI, 2009, pp. 186-187.

[11] Seo K, Wober M. Steinvurzel P, Schonbrun E, Dan Y, Ellenbogen T, Crozier K. B. Multicolored vertical silicon nanowires. Nano Letters. 2011;11:1851-1856. DOI: 10.1021/nl200201b.

[12] Yoon J.-S, Kim K, Meyyappan M, Baek C.-K. Optical characteristics of silicon-based asymmetric vertical nanowire photodetectors. IEEE Transactions on Electron Devices. 2017;64:2261-2266. DOI: 10.1109/TED.2017.2682878.

[13] Kwon H, Yoon J.-S, Lee Y, Kim D. Y, Baek C.-K, Kim J. K. An array of metal oxides nanoscale hetero p-n junctions toward designable and highly-scalable gas sensors. Sensors and Actuators B: Chemical. 2018;255:1663-1670. DOI: 10.1016/j.snb.2017.08.173.

[14] Lee Y, Kwon H, Yoon J.-S, Kim J. K. Overcoming ineffective resistance modulation in p-type NiO gas sensor by nanoscale Schottky contacts. Nanotechnology. 2019;30:115501-1-6. DOI: 10.1088/1361-6528/aaf957.

[15] Auth C, Aliyarukunju A, Asoro M, Bergstrom D, Bhagwat V, Birdsall J, Bisnik N, Buehler M, Chikarmane V, Ding G, Fu Q, Gomez H, Han W, Hanken D, Haran M, Hattendorf M, Heussner R, Hiramatsu H, Ho B, Jaloviar S, Jin I, Joshi S, Kirby S, Kosaraju S, Kothari H, Leatherman G, Lee K, Leib J, Madhavan A, Marla K, Meyer H, Mule T, Parker C, Parthasarathy S, Pelto C, Pipes L, Post I, Prince M, Rahman A, Rajamani S, Saha A, Dacuna Santos J, Sharma M, Sharma V, Shin J, Sinha P, Smith P, Sprinkle M, St. Amour A, Staus C, Suri R, Towner D, Tripathi A, Tura A, Ward C, Yeoh A. A 10nm high performance and low-power CMOS technology featuring 3rd generation FinFET transistors, self-aligned quad patterning, contact over active gate and cobalt local interconnects. In: Proceedings of 2017 IEEE International Electron Devices Meeting (IEDM), San Francisco, CA, 2017, pp. 29.1.1-29.1.4. DOI: 10.1109/IEDM.2017.8268472.

[16] Yeap G. Lin S. S, Chen Y. M, Shang H. L, Wang P. W, Lin H. C, Peng Y. C, Sheu J. Y, Wang M, Chen X, Yang B. R, Lin C. P, Yang F. C, Leung Y. K, Lin D. W, Chen C. P, Yu K. F, Chen D. H, Chang C. Y, Chen H. K, Hung P, Hou C. S, Cheng Y. K, Chang J, Yuan L, Lin C. K, Chen C. C, Yeo Y. C, Tsai M. H, Lin H. T, Chui C. O, Huang K. B, Chang W, Lin H. J, Chen K. W, Chen R, Sun S. H, Fu Q, Yang H. T, Chiang H. T, Yeh C. C, Lee T. L, Wang C. H, Shue S. L, Wu C. W, Lu R, Lin W. R, Wu J, Lai F, Wu Y. H, Tien B. Z, Huang Y. C, Lu L. C, He Jun, Ku Y, Lin J, Cao M, Chang T. S, Jang S. M. 5nm CMOS production technology platform featuring full-fledged EUV, and high mobility channel FinFETs with densest 0.021μm^2 SRAM cells for mobile SoC and high performance computing applications. In: Proceedings of 2019 IEEE International Electron Devices Meeting (IEDM), San Francisco, CA, USA, 2019, pp. 36.7.1-36.7.4. DOI: 10.1109/IEDM19573.2019.8993577.

[17] Yoon J.-S, Jeong J, Lee S, Baek R.-H. Systematic DC/AC performance benchmarking of sub-7-nm node FinFETs and nanosheet FETs. IEEE Journal of the Electron Devices Society. 2018;6: 942-947. DOI: 10.1109/JEDS.2018.2866026.

[18] Synopsys Inc., Mountain View, CA, Version O-2018.06, 2018.

[19] Yoon J.-S, Jeong J, Lee S, Baek R.-H. Multi-V_{th} strategies of 7-nm node nanosheet FETs with limited nanosheet spacing. IEEE Journal of the Electron Devices Society. 2018;6:861-865. DOI: 10.1109/JEDS.2018.2859799.

[20] Yoon J.-S, Jeong J, Lee S, Baek R.-H. Optimization of nanosheet number and width of multi-stacked nanosheet FETs for sub-7-nm node system on chip applications. Japanese Journal of Applied Physics. 2019;58:SBBA12-1-5. DOI: 10.7567/1347-4065/ab0277.

[21] Jeong J, Yoon J.-S, Lee S, Baek R.-H. Comprehensive analysis of source and drain recess depth variations on silicon nanosheet FETs for sub 5-nm node SoC application. IEEE Access. 2020;8:35873-35881. DOI: 10.1109/ACCESS.2020.2975017.

[22] Yoon J.-S, Jeong J, Lee S, Baek R.-H. Sensitivity of source/drain critical dimension variations for sub-5-nm node fin and nanosheet FETs. IEEE Transactions on Electron Devices. 2020;67:258-262. DOI: 10.1109/TED.2019.2951671.

[23] Yoon J.-S, Jeong J, Lee S, Baek R.-H. Punch-through-stopper free nanosheet FETs with crescent inner-spacer and isolated source/drain. IEEE Access. 2019;7:38593-38596. DOI: 10.1109/ACCESS.2019.2904944.

[24] Yoon J.-S, Jeong J, Lee S, Baek R.-H. Bottom oxide bulk FinFETs without punch-through-stopper for extending toward 5-nm node. IEEE Access. 2019;7: 75762-75767. DOI: 10.1109/ACCESS.2019.2920902.

[25] Wu H, Gluschenkov O, Tsutsui G, Niu C, Brew K, Durfee C, Prindle C, Kamineni V, Mochizuki S, Lavoie C, Nowak E, Liu Z, Yang J, Choi S, Demarest J, Yu L, Carr A, Wang W, Strane J, Tsai S, Liang Y, Amanapu H, Saraf I, Ryan K, Lie F, Kleemeier W, Choi K, Cave N, Yamashita T, Knorr A, Gupta D, Haran B, Guo D, Bu H, Khare M. Parasitic resistance reduction strategies for advanced CMOS FinFETs beyond 7nm. In: Proceedings of 2018 IEEE International Electron Devices Meeting (IEDM), San Francisco, CA, 2018, pp. 35.4.1-35.4.4. DOI: 10.1109/IEDM.2018.8614661.

[26] Yoon J.-S, Lee S, Lee J, Jeong J, Yun H, Kang B, Baek R.-H. Source/Drain patterning FinFETs as solution for physical area scaling toward 5-nm node. IEEE Access. 2019;7:172290-172295. DOI: 10.1109/ACCESS.2019.2956503.

[27] Yoon J.-S, Kim K, Rim T, Baek C.-K. Performance and variations induced by single interface trap of nanowire FETs at 7-nm node. IEEE Transactions on Electron Devices. 2017;64:339-345. DOI: 10.1109/TED.2016.2633970.

[28] International Roadmap for Devices and Systems (IRDS), 2020 Edition. Available from: https://irds.ieee.org/editions/2020.

[29] Bangsaruntip S, Cohen G. M, Majumdar A, Zhang Y, Engelmann S. U, Fuller N. C. M, Gignac L. M, Mittal S, Newbury J. S, Guillorn M, Barwicz T, Sekaric L, Frank M. M, Sleight J. W. High performance and highly uniform gate-all-around silicon nanowire MOSFETs with wire size dependent scaling. In: Proceedings of 2009 IEEE International Electron Devices Meeting (IEDM), Baltimore, MD, 2009, pp. 1-4. DOI: 10.1109/IEDM.2009.5424364.

[30] Na M. H, Nowak E. J, Haensch W, Cai J, The effective drive current in CMOS inverters. In: Proceedings of Digest. International Electron Devices Meeting, San Francisco, CA, USA, 2002, pp. 121-124. DOI: 10.1109/IEDM.2002.1175793.

[31] Baek R.-H, Baek C.-K, Jung S.-W, Yeoh Y. Y, Kim D.-W, Lee J.-S, Kim D. M, Jeong Y.-H. Characteristics of the series resistance extracted from Si nanowire FETs using the *Y*-function technique. IEEE Transactions on Nanotechnology. 2010;9:212-217. DOI: 10.1109/TNANO.2009.2028024.

[32] Zangenberg N. R, Fage-Pedersen J, Lundsgaard Hansen J, Nylandsted Larsen A. Boron and phosphorus diffusion in strained and relaxed Si and SiGe. Journal of Applied Physics. 2003;94:3883-3890. DOI: 10.1063/1.1602564.

[33] Jeong J, Yoon J.-S, Lee S, Baek R.-H. Threshold voltage variations induced by $Si_{1-x}Ge_x$ and $Si_{1-x}C_x$ of sub 5-nm node silicon nanosheet field-effect transistors. Journal of Nanoscience and Nanotechnology. 2020;20:4684-4689. DOI: 10.1166/jnn.2020.17799.

[34] Yoon J.-S, Lee S, Lee J, Jeong J, Yun H, Baek R.-H. Reduction of process variations for sub-5-nm node fin and nanosheet FETs using novel process scheme. EEE Transactions on Electron Devices. 2020;67:2732-2737. DOI: 10.1109/TED.2020.2995340.

[35] Oh Y.-S, Ward D. E. A calibrated model for trapping of implanted dopants at material interface during thermal annealing. In: Proceedings of International Electron Devices Meeting 1998. Technical Digest (Cat. No.98CH36217), San Francisco, CA, USA, 1998, pp. 509-512. DOI: 10.1109/IEDM.1998.746409.

Chapter 10

Engineering the Color and the Donor-Acceptor Behavior in Nanowires: Blend Versus Coaxial Geometry

Mohamed Mbarek and Kamel Alimi

Abstract

The blending or the bilayering of two complementary species are the dominant methods for in-solution-processed thin film devices to get a strong donor-acceptor behavior. They propose opposite strategies for the respective arrangement of the two species, a central point for energy and/or charge transfer. In this work, we propose to engineer at the scale of the exciton diffusion length the organization of a donor (poly(vinyl-carbazole), PVK) and an acceptor (poly(para-phenylene-vinylene), PPV) in a nanowire geometry. A two-step template strategy was used to fabricate coaxial nanowires with PPV and PVK, alternatively as the core or the shell material. Their stationary and time-resolved photoluminescence properties were investigated and compared to the case of PVK-PPV blend. Their respective characteristics are direct evidences of the dominant mechanisms responsible for the emission properties.

Keywords: organic nanowires, luminophores, exciton, shell-core

1. Introduction

π-conjugated molecules and polymers have attracted considerable interest in both fundamental studies and applied research [1]. This is due to their one-dimensional characteristics and their many potential applications. One-dimensional (1D) nanostructures based on organic materials are attracting significant research [2–4] interest owing to the many novel chemical, physical, and electronic properties that may arise in such systems and the possibility of exploiting these properties in a variety of applications [5–9]. However, for 1D nanostructure in general, a key challenge is the development of new approaches that will permit controlled nanowires and nanotubes architectures [10]. Nanoporous template method has been used to synthesize well-defined nanoscale tubes or wires inside nanoporous channels [10, 11]. This technique has many advantages, in that it is low-cost, control of physical properties or size is easy [10, 11], and that it lends itself to the logic of mass production. The fabrication of nanoscale structures has attracted much interest due to their potential use in electronic and optical applications [12–16]. Among many techniques generating the nanoscale structures, the anodized aluminum oxide (AAO) [10, 11] has been most widely used as a template for forming metal or semiconductor nanowires. The AAO has been known to have a clear advantage of an economical way to produce the large area periodic nanostructures [11]. These nanostructures were a seat of charge and energy transfer especially on the acceptor-donor architecture [17, 18]. The energy transfer [19] depend the absorption, the quantum yield of emitter's efficiency. Additionally, it depend on the donor and acceptor characters of luminophores, the relative spatial distribution of acceptor and donor, the overlapping between the donor absorption spectrum and the acceptor emission spectrum.

In this way, we report the synthesis of poly(para-phenylene-vinylene) PPV [20] (electron donor) and poly(vinyl-carbazole) PVK [20] (good hole transport) nanowires, and a coaxial architecture bases on PPV and PVK polymers using the AAO and describe their optical and structural characteristics. Stationary and transition photoluminescence have been employed to demonstrate the nature of energy and charge-transfer processes occurring in these nanostructures.

2. Materials and techniques

2.1 Materials

Poly(N-vinylcarbazole) PVK powder, chloroform, methanol, used for the synthesis of the studied compounds were purchased from Sigma Aldrich, Merk, and Fluka. The materials were used as received. The PPV precursor was synthesized by addition of 33 mL of tetrahydrothiophenium in the dichloroparaxylene dissolved in methanol. The aqueous solution of PPV precursor was dialyzed with deionized water for several days. The PPV precursor was kept at 0°C in the dark. The mass concentration of the PPV precursor is about 2.4 mg mL. Commercial AAO membranes purchased from Whatman (anodisc 13) have been used. They are 60 μm thick with a real pore diameter showing a polydispersity between200 nm and around 250 nm as revealed by SEM analysis.

2.2 Fabrication of the coaxial NWs

The coaxial NWs were fabricated by a wetting template method in anodic alumina oxide (AAO) nanoporous template. For the synthesis of coaxial NWs, two steps of wetting are required. The first step consists in the wetting of AAO membrane with the PPV precursor in methanol solution. To obtain PPV nanotubes, the concentration of PPV precursor in methanol was chosen at 1.2 mg/mL and 200 μL of this solution was drop-casted on AAO membrane, as described elsewhere. The precursor embedded in AAO membrane was then thermally converted under a dynamic secondary vacuum (≈10–6 Torr) for 4 h to obtain PPV. The coaxial nanowires are then obtained by a secondary-template strategy. The AAO membrane containing PPV nanotubes was wetted with the solution containing PVK dispersed in chloroform. The wetted templates were left overnight under ambient condition to allow solvent evaporation. For some characterization, the AAO templates containing coaxial NWs was completely etched in H3PO4 (25 wt %) overnight and washed several time with DI water. The coaxial NWs were homogeneously dispersed in DI water solution by ultrasonication during 10 s with a power of 140 W (Fisher Scientific FB 15052). Ultrasonication results in the shortening of the nanowire length.

3. Characterization methods

A field-effect scanning electron microscope (JEOL, JSM-7600F operating at 5 kV), an AFM (Nanowizard II, JPK instruments) working in intermittent contact mode in air with Si tips (PPP-NCHR, Nanosensors) and a transmission electron microscope (Hitachi H9000 NAR operating at 300 kV) were used to investigate the morphology and the composition of the nanowires. For SEM, AFM and fluorescence microscopy experiments, a drop of solution (10 μL) containing the nanowires was deposited onto silicon or glass substrates after the dissolution of the template. For TEM experiments, a drop of the solution was deposited on TEM copper grids covered with a thin holeycarbon film. Epi-fluorescence micrographs were acquired using a calibrated microscope (Eclipse Ti, Nikon) equipped with a 60 objective and a CCD camera, 130WHg lamp and fluorescence filter cube (EX 330–380, Dm: 400, BA 420). Spectroscopic Characterizations.

All characterizations were performed at room temperature. Absorption and photoluminescence of the PVK were measured on spin coated thin films (1500 rpm, 30 s) deposited on glass and silica from a solution with 2.4 mg/mL-1, and from a PPV precursor (PPV precursor: 2.4 mg/mL-1 in methanol, 1500 rpm, 30 s, thermal conversion at 300°C for 3 h). UV–vis absorption on thin film sample was performed with a Perkin-Elmer double beam spectrophotometer equipped with an integrating sphere; we carried out PL measurements using a Jobin-Yvon Fluorolog 3 equipped with a CCD camera and PL experiments on film were achieved under 400 nm excitation at 0.5 mW by a Spectra-Physics Hurricane X laser. For micro-Raman and micro photoluminescence studies, the template containing the nanowires was broken to reveal a cross section and consequently the whole length of the nanowires. Micro-Raman spectra were recorded using a Renishaw inVia Raman microscope equipped with a 785 nm line of a HPNIR diode laser. We measured steady state micro-PL spectra with a Jobin-Yvon T64000 spectrometer under 325 nm laser excitation obtained by an argon ion laser. In each case, the spot size of the focused laser beam on the sample was estimated to be about 1.2 μm. The transient PL experiments have been achieved under 400 nm excitation using Spectra-Physics Hurricane X laser system (82 fs, 1 kHz) onto mat of NWs. The collected emission was temporally detected with a streak camera (Hamamatsu C7700) coupled with an imaging spectrograph. The laser pump power in pinging on sample was kept at 0.5 mW to minimize sample photobleaching.

4. Results and discussion

A selection of transmission electron microscopy (TEM) was performed to characterize the different nanowires based on PPV and PVK. The different types of TEM imaging and their correspondent's histograms are shown in the **Figure 1**.

A selection of transmission electron microscope (TEM) images of the as-prepared nanofibres after the dissolution of the AAO membrane is presented in **Figure 1**. The PPV nanofibres have a mean diameter d (272 nm) exceeding the nominal manufacturer pore diameter dp (200 nm) of AAO membrane as shown by the histograms. The same diameter was observed in the PVK nanowires displayed in TEM micrographs with a mean diameter d = 270 nm (**Figure 1b**) supported by their histograms nanowires diameters in the right of **Figure 1**.The nanowires length 10 μm does not correspond to the membrane thickness (60 μm), this difference can be explained by the alumina removal step and the sonication step required for their dispersion in solution, which result in shortening of the NWs.

We performed also transmission electron microscopy (TEM) and scanning electron microscopy experiments on the coaxial nanowires to make evidence of the coaxial architecture and to prove the successful synthesis of these coaxial nanostructures bases on PVK and PPV polymers. The poor imaging contrast typical of polymeric materials make not easy to prove that. In **Figure 2**.

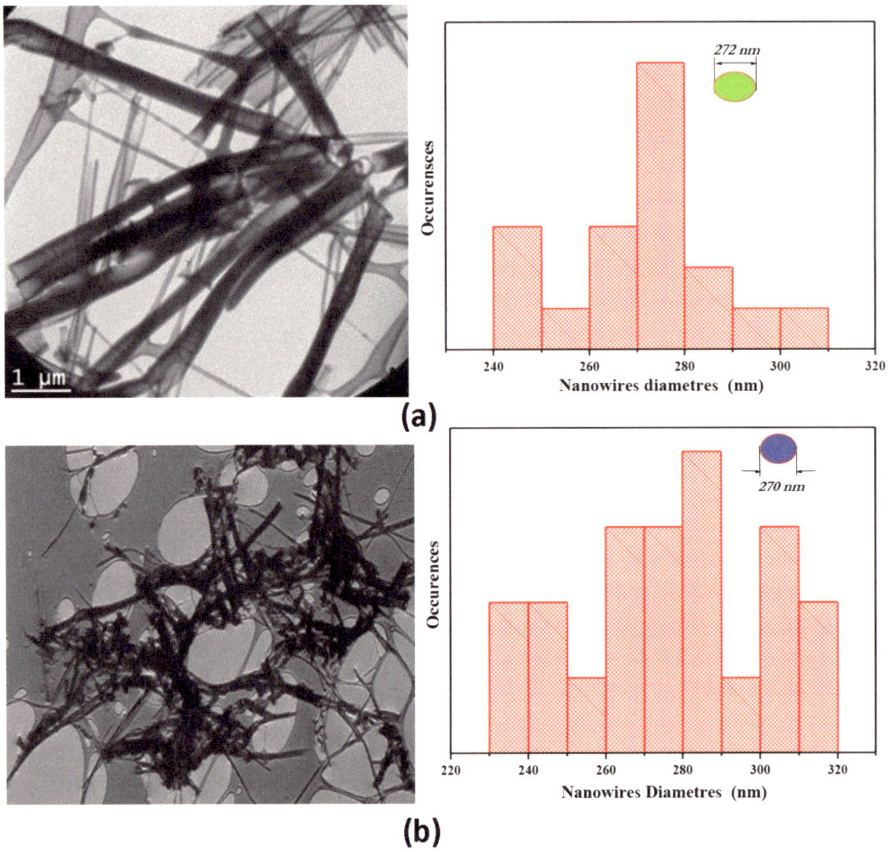

Figure 1.
Selection of TEM micrographs of: (a) PPV nanowires and it histograms nanowires diameters (right of TEM image), (b) of PVK nanowires and it histograms nanowires diameters (right of TEM micrograph).

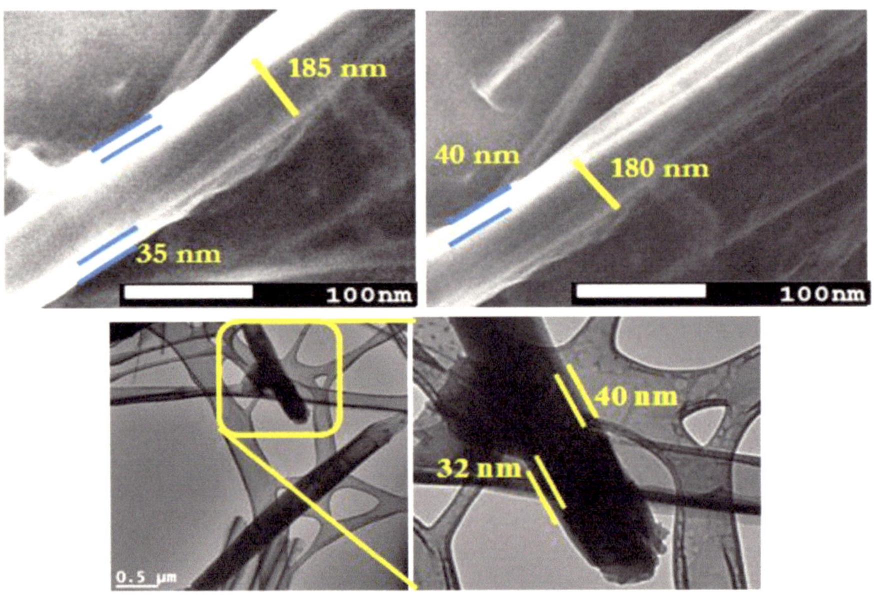

Figure 2.
Selection of SEM and TEM images of coaxial nanowires based on PVK and PPV polymers of an isolated NW which reveals the coaxial geometry.

Generally, this strategy promotes complex phenomena occurring when two (or more) kinds of emitters are into contact, that are Dexter and Förster energy transfers as well as charge transfers. The first mechanism is the energy transfer between PVK and PPV. The emission spectrum of PVK has some overlap with the absorption spectrum of PPV. The spectral overlap is large enough for the efficient energy transfer according to the Forster theory [21]. The energy transfer process from PVK to PPV enhances the PL intensity of PPV.

Here, an energy transfer was accrued between two luminophores because the absorption band of the PPV unit (green dashed-dot line, peak at 420 nm) has a significant overlap with the emission band of the PVK (blue line, peak at 400 nm) as shown on **Figure 3**. Thus, the absorption by the PPV units of the photons emitted by PVK can significantly occur and make evidence of reabsorption phenomenon (energy transfer). This phenomenon was more clearly visible in the PL spectra as shown in **Figure 4**.

The emissive characteristics of the as-prepared NWs have been investigated by PL spectroscopy with a 2 μm diameter microprobe for pumping at 325 nm along the membrane pores, as show in **Figure 4**. The bleu spectrum is composed of the characteristic PVK PL bands located at 389, 418 and 453 nm [20]. However in the case of the green spectra four bands which constitute the spectrum of the PPV are located at 520 nm, 550, 600 and 650 nm [20]. The obtained spectra of the mixture of [PVK@PPV] traced with magenta lines present a linear combination of the PVK and PPV spectra with a strongest contribution of PPV emission. In fact the intensity of PPV luminophores is triple of PVK emission intensity. It is clearly seen that the intensity of PVK is more than PPV emission intensity of about 1.5 before combined together. After mixing the two luminophores [PVK@PPV] the PPV emission intensity become triple of PVK emission intensity. So adding a type of luminophore witch her emission is overlapping with the absorption of the other luminophore resulting increase in emission of this latter.

The sum of two photoluminescence spectra is accompanied by a strong energy transfer as well as evidence of charge transfer. The emission energy of PVK was reabsorbed by PPV units which amplified the PPV emission in the mixture [PVK@PPV]. In addition, as previously reported the PVK emission is overlap of PPV absorption and this is leads to a Forster energy transfer [22–26]. In the same regard, we recall that the quantum yield of PL of PPV is more than of PVK. That means that the PPV polymer has more power to create emissive excitons than in

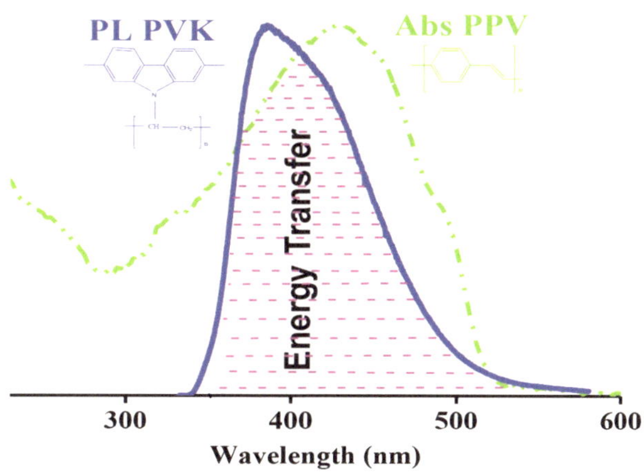

Figure 3.
Normalized absorption (dashed-dot lines), and photoluminescence (solid lines, λexc = 330 nm) spectra of the PVK (blue lines) and of the PPV polymer (green lines).

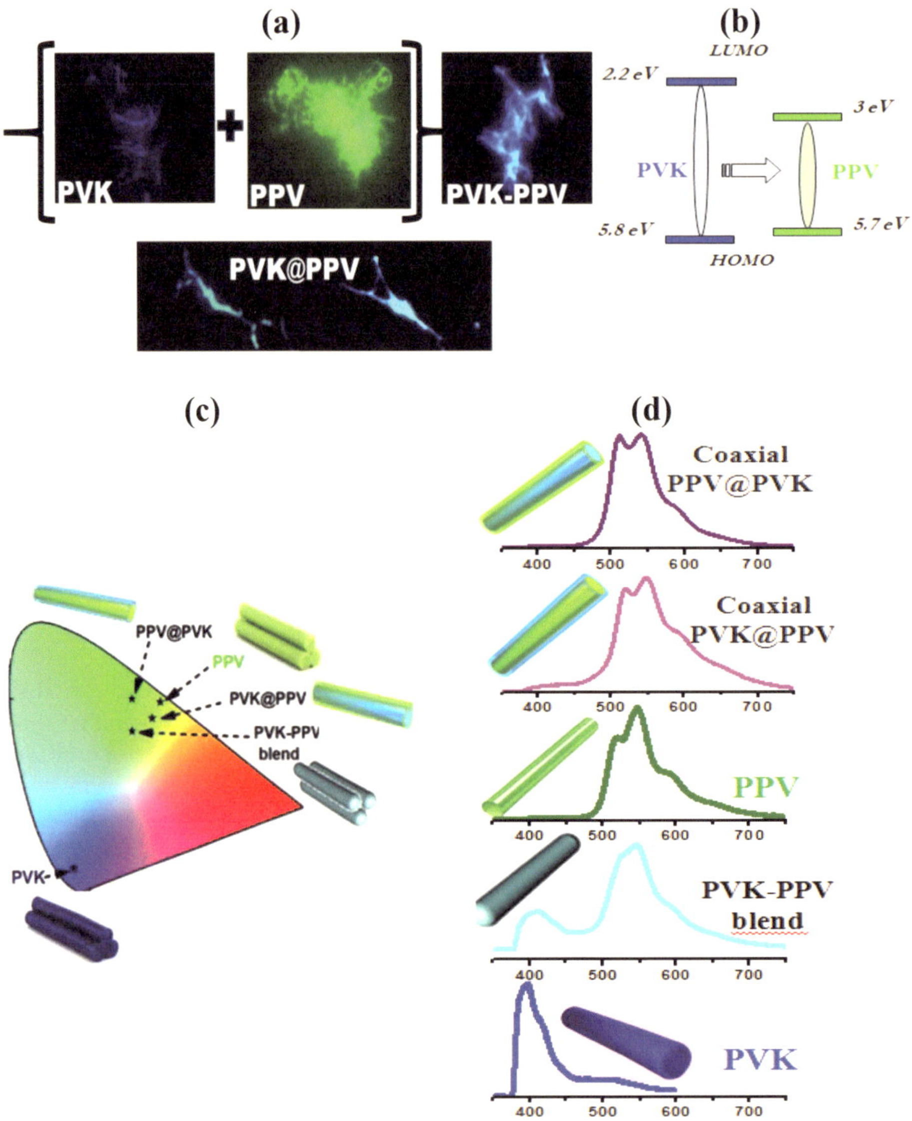

Figure 4.
(a) Fluorescence microscopy images (λexc = 330–380 nm; λem > 420 nm) of dispersed PPV nanowires (green), PVK nanowires (blue) and [PPV@PVK] nanowires (green-bleu) (b) energy transfer phenomena between PVK and PPV (HOMO: Highest occupied molecular orbital, LUMO: Lowest unoccupied molecular orbital); (c) chromaticity diagram (CIE coordinates) for PVK (x = 0.1575; y = 0.0814), PPV (x = 0.3787; y = 0.5919) and [PVK@PPV] (x = 0.3091; y = 0.4968) nanowires LEDs described; (d) photoluminescence (solid lines, laser λexc = 325 nm) spectra of the PVK nanowires (blue lines), PPV nanowires (green lines) and [PPV@PVK] nanowires (magenta lines).

the case of PVK. This is evidence of charge transfer between PPV and PVK units. For charge transfer phenomenon, a similar distance between the donor and the acceptor is needed. Concerning Förster energy transfer, the process is efficient for typical distances 5–10 nm between the donor and the acceptor [26].

Figure 4 shows epi-fluorescence micrographs of random PVK NWs, PPV NWs [PVK@PPV] blend nanowires and coaxial nanowires. A green emission of PPV, blue emission and a bright blue-green characteristic of PPV and PVK emissions are observed in the case of mixture [PVK@PPV]. The color emission corresponding to each PL spectrum has been determined for PVK, PPV and [PVK@PPV] nanowires and reported in a chromaticity diagram. The coordinates of PVK NWs (x = 0.1575; y = 0.0814), PPV NWs (x = 0.3787; y = 0.5919) and [PVK@PPV] NWs LEDs are displayed in **Figure 4**. It can be noted that the perception of the emission color for the mixture NWs [PVK@PPV] was dominated by the PPV one. Moreover, the contribution of the PPV emission is favored by the higher quantum yield of PPV.

The effect of the coaxial morphology on energy transfer dynamics is now examined. It has been investigated by time-resolved photoluminescence of PVK and PPV thin films, blended and coaxial PVK/PPV nanowires for an excitation at 3.10 eV (400 nm). The nanowires were dispersed on a silica substrate after the total etching of the template. Furthermore, the energy transfer dynamics was investigated by time resolving the PVK, PPV and [PVK@PPV] nanowires optical emission upon excitation at 3.10 eV (400 nm).

Due to the very weak signal measured in the spectral range of PVK emission and of the proximity of the excitation at 400 nm, it was not possible to address the dynamics of the main photoluminescence band. So, we focused on the emission at 535 nm, because it corresponds to the second emission band in the PL tail of PVK, and it is close to the PL maximum of PPV spectrum. Its dynamic behavior is thus expected to take into account the behavior of each polymer and an eventual donor-acceptor mechanism.

Figure 5 depicts the 535 nm PL intensity decays of all the systems investigated in this study. These PL kinetics are simulated with two exponential decays with characteristic time τ1 and τ2. In this simple model, the populations of levels 1 and 2 are independent. These populations include photogenerated charges recombining radiatively and non-radiatively. The decaying population is $n = A_1 n_1 + A_2 n_2$, where A1 and A2 are proportional to the PL intensity from levels 1 and 2, respectively [27]. The intensity averaged decay time τ_{mean} has been determined in order to show the average trend of the photogenerated charge migration time. The corresponding decay times and corresponding yields Pi defined as $P_i(\%) = A_i\tau_i / \sum_{i=1}^{2} A_i\tau_i$ are compiled in the **Table 1**.

A first analysis shows that the decays for the blend and the core-shell nanowires are in-between the decay of pure PPV (the faster) and PVK (the slower) decays. The corresponding average decay times are longer (183, 158 and 176 ps) than for

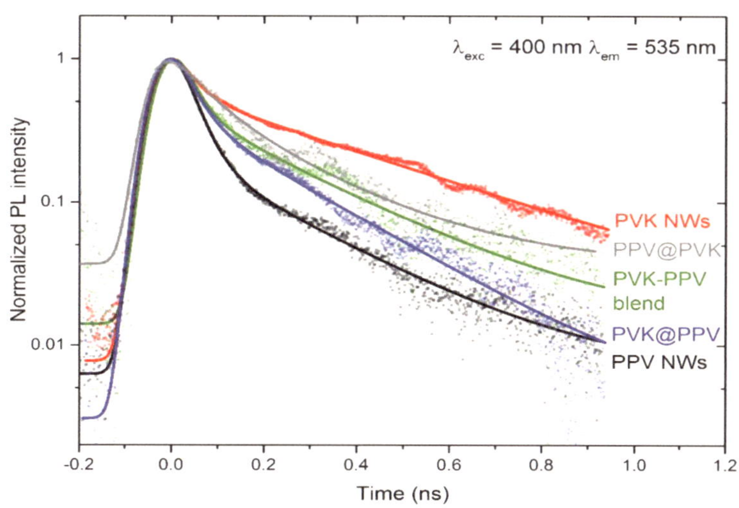

Figure 5.
Normalized photoluminescence decay curves of PVK (red) and PPV (black) nanowires, PVK-PPV blend nanowires (green), and coaxial PPV@PVK (gray) and PVK@PPV (blue) nanowires, extracted for PL emission intensity at 535 nm (excitation wavelength: 400 nm).

Nanowire material	τ_1[a] (ps)	τ_2[b] (ps)	P_1[c] (%)	P_2[d] (%)	τ_{mean}[e] (ps)
PPV	35	238	0.59	0.41	118
PVK	37	405	0.17	0.83	342
PPV-PVK blend	33	254	0.32	0.68	183
PVK@PPV	35	230	0.37	0.63	158
PPV@PVK	67	221	0.29	0.71	176

[a]*Lifetime of the photo-generated charge in the level 1.*
[b]*Lifetime of the photo-generated charge in the level 1.*
[c]*Weight relative of population of photo-generated charges contributing to level 1.*
[d]*Weight relative of population of photo-generated charges contributing to level 1.*
[e]*Average decay time.*

Table 1.
Photoluminescence decays times (in picosecond, with error bars estimated at about 10%) and yields of each component determined from the PL decays shown in ***Figure*** *5.*

pristine PPV (τ_{mean} = 118 ps) and shorter than for pristine PVK (τ_{mean} = 342 ps). It can be noted that at an emission wavelength of 535 nm, whatever the kind of nanowires, blend or coaxial, the PL of the PPV component dominates because PPV has the shortest lifetime and a quantum yield slightly larger. In one hand, the system which shows the decay closer to PVK is PPV@PVK. Its longer average lifetime can be attributed to the photogenerated excitons in the PVK when considered as a donor. They have an important probability to diffuse to the acceptor PPV and to transfer their energy before any radiative recombination. In another hand, the system which shows the decay closer to PPV is PVK@PPV. It suggests that the emission by the PPV shell is due to the excitation of PPV excitons partly by the laser probe, and partly by the energy transfer from the photogenerated excitons in the PVK acting as a donor. The case of PVK-PPV blend decay is intermediate between the two coaxial configurations, a bit closer from the PVK@PPV arrangement. It suggests that for this blend, PPV and PVK segments do not deeply affect the exciton dynamics of each other, as expected when chains of both polymers are not entangled, or at least close enough.

These results strongly support the analysis and the conclusion of the steady-state PL study, dominated by morphological issues in relation with the exciton diffusion length of PVK: thin PVK shell for the PPV@PVK nanowires, large PVK diameter for PVK@PPV nanowires, large PVK domains due to a poor intermixing between PPV and PVK obtained during the blending process. This work opens the way to develop alternative solution processing techniques to manage the local organization of donor-acceptor systems at the scale of the exciton diffusion length.

5. Conclusion

In this work, the template method with in-solution processes was exploited to control the respective organization of two polymers at the nanoscale in order to tune the donor-acceptor behavior, and then the photoluminescence properties of such nanowires. Two luminophores, PPV and PVK, were used for the fabrication of coaxial architectures and blend arrangement. The analysis of both steady-state and time-resolved photoluminescence study were analyzed by comparing the exciton diffusion length to the domain size of the nanowire shell, the nanowire core, or the segregated domains in the blend case. Charge and energy transfer process occurred in these coaxial NWs are more described using stationary and time resolving photoluminescence. These coaxial morphology based to PVK and PPV polymers showed an amplified of emission PPV intensity by adding PVK.

Acknowledgements

The authors would like to thank these colleagues at the Institute of Materials Jean Rouxel in Nantes: Jean-Luc duvail, Florian Massuyeau and Eric Faulques for help in optical study.

Author details

Mohamed Mbarek* and Kamel Alimi
Laboratory Research: Assymetric Synthesis and Molecular Engineering of New Materials for Organic Electronic (LR18ES19), Faculty of Sciences Monastir, Tunisia

*Address all correspondence to: mohamedmbarek99@yahoo.fr

References

[1] Choi M, Kim Y, Ha C. Polymers for flexible displays: From material selection to device applications. Prog. Polym. Sci. 2008;33:581-630 . https://doi.org/10.1016/j.progpolymsci.2007.11.004

[2] Chen S, Li Y, Li Y,Architecture of low dimensional nanostructures based on conjugated polymers. Polym Chem. 2013 ;4 :5162-5180 https://doi.org/10.1039/C3PY00098B

[3] Shane M, Pierre L, Deirdre O'C, Daniela I, Gareth R. Alignment and Dynamic Manipulation of Conjugated Polymer Nanowires in Nematic Liquid Crystal Hosts. Advanced Materials 2008;20:2497-2502. https://doi.org/10.1002/adma.200701986

[4] Lang J, Huanli D, Wenping H. Controlled growth and assembly of one-dimensional ordered nanostructures of organic functional materials Soft Matter, 2011;7:1615-1630

[5] Li X, Wang T, Zhang J, Zhu D, Zhang X, Ning Y, Zhang H, Yang B. Controlled Fabrication of Fluorescent Barcode Nanorods. ACS Nano 2010;4: 4350-4360. https://doi.org/10.1039/C0SM00762E

[6] Park D.H, Hong Y.K, Cho E.H, Kim M.S, Kim D.C, Bang J, Kim J, Joo J, Light-Emitting Color Barcode Nanowires Using Polymers: Nanoscale Optical Characteristics. ACS Nano 2010;4:5155-5162. https://doi.org/10.1021/nn101096m

[7] Gu F, Zhang L, Yin X, Tong L. Polymer Single-Nanowire Optical Sensors. Nano Lett. 2008;8:2757-2761. https://doi.org/10.1021/nl8012314

[8] Zhao Y.S, Wu J, Huang J. Vertical Organic Nanowire Arrays: Controlled Synthesis and Chemical Sensors. J. Am. Chem. Soc. 2009;131:3158-3159. https://doi.org/10.1021/ja809360v

[9] Zang L, Che Y, Moore J.S. One-Dimensional Self- Assembly of Planar pi-Conjugated Molecules: Adaptable Building Blocks for Organic Nanodevices. Acc. Chem. Res. 2008;41:1596-1608. https://doi.org/10.1021/ar800030w

[10] Mbarek M, Garreau A, Massuyeau F, Alimi K, Wéry J, Faulques E, Duvail. J.L. Template process for engineering the photoluminescence of PVK and PPV-based nanowires". Journal of Apllied Polymer Sciences 2019;136:48201. https://doi.org/10.1002/app.48201

[11] Garreau A, Massuyeau F, Cordier S, Molard Y, Gautron E, Bertoncini P, Faulques E, Wery J, Humbert B, Bulou A, Duvail J-L. Color Control in Coaxial Two-Luminophore Nanowires ACS Nano 2013 ;7:2977-2987. https://doi.org/10.1021/nn400763r

[12] Che Y, Yang X, Loser S, Zang L. Expedient Vapor Probing of Organic Amines Using Fluorescent Nanofibers Fabricated from an n-Type Organic Semiconductor. Nano Lett. 2008;8:2219-2223. https://doi.org/10.1021/nl080761g

[13] O'Carroll D, Lieberwirth I, Redmond,G. Microcavity Effects and Optically Pumped Lasing in Single Conjugated Polymer Nanowires. Nat. Nanotechnol. 2007;2:180-184. https://doi.org/10.1038/nnano.2007.35

[14] Zhao Y.S, Peng A, Fu H, Ma Y, Yao J. Nanowire Waveguides and Ultraviolet Lasers Based on Small Organic Molecules. Adv. Mater. 2008;20: 1661-1665. https://doi.org/10.1002/adma.200800123

[15] Kuo C.C, Lin C.H, Chen W.C. Morphology and Photophysical Properties of Light-Emitting Electrospun Nanofibers Prepared from Poly(fluorene) Derivative/PMMA Blends. Macromolecules

2007;40:6959-6966. https://doi.org/10.1021/ma071182l

[16] Yang H, Lightner C.R, Dong L. Light-Emitting Coaxial Nanofibers. ACS Nano 2012;6:622-628. https://doi.org/10.1021/nn204055t

[17] Zhao Y.S, Fu H.B, Hu F.Q, Peng A.D, Yang W.S, Yao J.N. Tunable Emission from Binary Organic One-Dimensional Nanomaterials: An Alternative Approach to White-Light Emission. Adv. Mater. 2008;20:79-83. https://doi.org/10.1002/adma.200890030

[18] Simbrunner C, Hernandez-Sosa G, Quochi F, Schwabegger G.N, Botta C, Oehzelt M, Salzmann I, Djuric T, Neuhold A, Resel R, Color Tuning of Nanofibers by Periodic Organic-Organic Hetero-Epitaxy. ACS Nano 2012;6:4629-4638. https://doi.org/10.1021/nn2047235

[19] Lakowicz J.R, Principles of Fluorescence Spectroscopy, 3rd ed.; Springer: New York, 2006. https://www.springer.com/gp/book/9780387312781

[20] Mbarek M, Massuyeau F, Duvail J.L, Wéry J, Faulques E, Alimi K. New copolymer of poly (N-vinylcarbazole) and poly (p-phenylenevinylene) for optoelectronic devices. Journal of Apllied Polymer Sciences 2013;130:2839-2847), DOI: 10.1002/app.39496.

[21] Chi-Jung C, Yao-Yi C, Min Huei W, Chi Shen T. Miscibility effect on the energy transfer induced luminescence enhancement of polyimide/PVK blends Thin Solid Films 2005;477:14-18. https://doi.org/10.1016/j.tsf.2004.08.104

[22] Dennis A.M, Bao G. Quantum Dot-Fluorescent Protein Pairs as Novel Fluorescence Resonance Energy TransferProbes. Nano Lett. 2008;8:1439-1445. https://doi.org/10.1021/nl080358+

[23] Stoferle T, Scherf U, Mahrt R.F. Energy Transfer in Hybrid Organic/Inorganic Nanocomposites. Nano Lett. 2009;9:453-456. https://doi.org/10.1021/nl8034465

[24] Bredas J.L, Beljonne D, Coropceanu V, Cornil J. Charge- Transfer and Energy-Transfer Processes in pi-Conjugated Oligomers and Polymers: A Molecular Picture. Chem. Rev. 2004;104:4971-5003. https://doi.org/10.1021/cr040084k

[25] Forster, T. Transfer Mechanisms of Electronic Excitation. Discuss. Faraday Soc. 1959;27:7-17. https://doi.org/10.1039/DF9592700007

[26] Halivni S, Sitt A, Hadar I, Banin U, Effect of Nanoparticle Dimensionality on Fluorescence Resonance Energy Transfer in Nanoparticle-Dye Conjugated Systems. ACS Nano 2012;6:2758-2765. https://doi.org/10.1021/nn300216v

[27] Mohamed M, Lamia S, Kamel A, New copolymer involving PVK and F8BT for organic solar cells applications: Design, synthesis, characterization and theoretical studies". Optical Materials 2019;91:447-454. https://doi.org/10.1016/j.optmat.2019.03.053